AS/A2 Level Chemistry courses

AS and A2

All Chemistry A Level courses currently studied are in two parts, with three separate modules in each part. Students first study the AS (Advanced Subsidiary) course. Some will then go on to study the second part of the A Level course, called A2. Advanced Subsidiary is assessed at the standard expected halfway through an A Level course: i.e., between GCSE and Advanced GCE. This means that new AS and A2 courses are designed so that difficulty steadily increases:

- AS Chemistry builds from GCSE science
- A2 Chemistry builds from AS Chemistry.

How will you be tested?

Assessment units

For AS Chemistry, you will be tested by three assessment units. For the full A Level in Chemistry, you will take a further three units. AS Chemistry forms 50% of the assessment weighting for the full A Level.

One of the units in AS and in A2 is practically based. You will take two theory units in each of AS and A2. Each theory unit can normally be taken in either January or June. Alternatively, you can study the whole course before taking any of the unit tests.

If you are disappointed with a module result, you can resit each module. The higher mark counts.

Coursework

Coursework forms part of your A Level Chemistry course in the form of the assessment of practical skills. More details are provided on page 6.

Key skills

It is important that you develop your key skills of Communication, Application of Number and Information Technology throughout your AS and A2 courses. These are important skills that you need whatever you do beyond AS and A Levels. To gain the key skills qualification, you will need to demonstrate your ability to put your ideas across to other people, collect data and use up-to-date technology in your work. You will have many opportunities during A2 Chemistry to develop your key skills.

AS/A2 Level Chemistry courses

What skills will I need?

For A2 Chemistry, you will be tested by *assessment objectives*: these are the skills and abilities that you should have acquired by studying the course. The assessment objectives for A2 Chemistry are shown below.

Knowledge and understanding of science and How Science Works

- recognising, recalling and showing understanding of scientific knowledge
- selection, organisation and communication of relevant information in a variety of forms.

Application of knowledge and understanding of science and How Science Works

- analysis and evaluation of scientific knowledge and processes
- applying scientific knowledge and processes to unfamiliar situations including those related to issues
- assessing the validity, reliability and credibility of scientific information.

How Science Works

- demonstrating and describing ethical, safe and skilful practical techniques and processes
- selecting appropriate qualitative and quantitative methods
- making, recording and communicating reliable and valid observations and measurements with appropriate precision and accuracy
- analysis, interpretation, explanation and evaluation of the methodology, results and impact of their own and others' experimental and investigative activities in a variety of ways.

Experimental and investigative skills

Chemistry is a practical subject and part of the assessment of A2 Chemistry will test your practical skills. This will be carried out during your lessons. You will be assessed on three main skills:

- implementing
- analysing evidence and drawing conclusions
- evaluating evidence and procedures.

The skills may be assessed in separate practical exercises. They may also be assessed all together in the context of a single 'whole investigation'.

You will receive guidance about how your practical skills will be assessed from your teacher. This Study Guide concentrates on preparing you for the written examinations testing the subject content of A2 Chemistry.

Synoptic assessment

- bringing together knowledge, principles and concepts from different areas of Chemistry and applying them in a particular context
- using chemical skills in contexts which bring together different areas of the subject.

More details of synoptic assessment in Chemistry are discussed in Chapter 6 of this Study Guide (pages 141–142).

Revise A2

AQA Chemistry

Contents

Specification list	4
AS/A2 Level Chemistry courses	5
Different types of questions in A2 examinations	7
Exam technique	9
What grade do you want?	11
Four steps to successful revision	12

Chapter 1 Rates and equilibria

1.1 Orders and the rate equation	13
1.2 Determination of reaction mechanisms	17
1.3 The equilibrium constant, K_c	19
1.4 Acids and bases	24
1.5 The pH scale	27
1.6 pH changes	31
Sample question and model answer	36
Practice examination questions	37

Chapter 2 Energy changes in chemistry

2.1 Enthalpy changes	39
2.2 Entropy changes	45
2.3 Electrochemical cells	48
2.4 Predicting redox reactions	54
Sample question and model answer	56
Practice examination questions	57

Chapter 3 The Periodic Table

3.1 Chemistry of the Period 3 elements	59
3.2 The transition elements	63
3.3 Transition element complexes	66
3.4 Ligand substitution of complex ions	71
3.5 Redox reactions of transition metal ions	74
3.6 Catalysis	77
3.7 Reactions of metal aqua-ions	80
Sample question and model answer	84
Practice examination questions	85

Chapter 4 Chemistry of organic functional groups

4.1	Isomerism and functional groups	87
4.2	Aldehydes and ketones	90
4.3	Carboxylic acids	94
4.4	Esters	97
4.5	Acylation	100
4.6	Arenes	103
4.7	Reactions of arenes	107
4.8	Amines	109
	Sample question and model answer	113
	Practice examination questions	114

Chapter 5 Polymers, analysis and synthesis

5.1	Amino acids	116
5.2	Polymers, proteins and nucleic acids	120
5.3	Chromatography	125
5.4	Analysis	126
5.5	NMR spectroscopy	130
5.6	Organic synthetic routes	136
	Sample question and model answer	138
	Practice examination questions	139

Chapter 6 Synoptic assessment

What is synoptic assessment?	141
Worked synoptic sample question	142
Practice examination questions	143

Practice examination answers	145
Periodic table	158
Index	159

Specification list

AQA A2 Chemistry

MODULE	SPECIFICATION TOPIC	CHAPTER REFERENCE	STUDIED IN CLASS	REVISED	PRACTICE QUESTIONS
A2 Unit 1 (M4) Kinetics, equilibria and organic chemistry	Kinetics	1.1, 1.2			
	Equilibria	1.3			
	Acids and bases	1.4, 1.5, 1.6			
	Nomenclature and isomerism on organic chemistry	4.1			
	Compounds containing the carbonyl group	4.2, 4.3, 4.4, 4.5			
	Aromatic chemistry	4.6, 4.7			
	Amines	4.8			
	Amino acids	5.1, 5.3			
	Polymers	5.2			
	Organic synthesis and analysis	5.6			
	Structural determination	5.3, 5.4, 5.5			
A2 Unit 2 (M5) Energetics, redox and inorganic chemistry	Thermodynamics	2.1, 2.2, AS 3.2			
	Periodicity	3.1			
	Redox equilibria	2.3, 2.4			
	Transition metals	3.2, 3.3, 3.5, 3.6			
	Reactions of inorganic compounds in aqueous solution	3.4, 3.7			

Examination analysis

A2 Chemistry comprises two unit tests. All questions are compulsory.
Synoptic assessment will form part of both units. Practical and investigative skills will also be assessed.

Units 1, 2 and 3 comprise AS Chemistry		50%
Unit 4 Structured questions: short and extended answers	1 hr 45 min test	20%
Unit 5 Structured questions: short and extended answers	1 hr 45 min test	20%
Unit 6 Internal assessment of practical and investigative skills		10%

Different types of questions in A2 examinations

In A2 Chemistry examinations, different types of question are used to assess your abilities and skills. Unit tests mainly use structured questions requiring both short answers and more extended answers.

Short-answer questions

A short-answer question may test recall or it may test understanding. Short answer questions normally have space for the answers printed on the question paper.

Here are some examples (the answers are shown in blue):

What is meant by isotopes?

Atoms of the same element with different mass numbers.

Calculate the amount (in mol) of H_2O in 4.5 g of H_2O.

1 mol H_2O has a mass of 18 g. ∴ 4.5 g of H_2O contains 4.5/18 = 0.25 mol H_2O.

Structured questions

Structured questions are in several parts. The parts usually have a common context and they often become progressively more difficult and more demanding as you work your way through the question. A structured question may start with simple recall, then test understanding of a familiar or an unfamiliar situation.

Here is an example of a structured question that becomes progressively more demanding.

(a) Write down the atomic structure of the two isotopes of potassium: ^{39}K and ^{41}K.

  (i) ^{39}K ...19... protons; ...20... neutrons; ...19... electrons. ✓
  (ii) ^{41}K ...19... protons; ...22... neutrons; ...19... electrons. ✓ [2]

(b) A sample of potassium has the following percentage composition by mass:
^{39}K: 92%; ^{41}K: 8% Calculate the relative atomic mass of the potassium sample.

  92 × 39/100 + 8 × 41/100 = 39.16 ✓ [1]

(c) What is the electronic configuration of a potassium atom?

  $1s^2 2s^2 2p^6 3s^2 3p^6 4s^1$ ✓ [1]

(d) The second ionisation energy of potassium is much larger than its first ionisation energy.

  (i) Explain what is meant by the *first ionisation energy* of potassium.

  The energy required to remove an electron ✓ from each atom in 1 mole ✓ of gaseous atoms ✓

  (ii) Why is there a large difference between the values for the first and the second ionisation energies of potassium?

  The 2nd electron removed is from a different shell ✓ which is closer to the nucleus and experiences more attraction from the nucleus. ✓ This outermost electron experiences less shielding from the nucleus because there are fewer inner electron shells than for the 1st ionisation energy. ✓ [6]

Different types of questions in A2 examinations

Extended answers

In A2 Level Chemistry, questions requiring more extended answers may form part of structured questions or may form separate questions. They may appear anywhere on the paper and will typically have between 5 and 15 marks allocated to the answers as well as several lines of answer space. These questions are also often used to assess your abilities to communicate ideas and put together a logical argument.

The correct answers to extended questions are often less well-defined than to those requiring short answers. Examiners may have a list of points for which credit is awarded up to the maximum for the question.

An example of a question requiring an extended answer is shown below.

> Choose two complex ions of copper with different shapes.
> Describe the shapes of, and bond angles in, these complex ions
> and explain what is meant by ligand substitution. [6 marks]

In aqueous condition, Cu^{2+} forms $[Cu(H_2O)_6]^{2+}$ ✓ complex ions. $[Cu(H_2O)_6]^{2+}$ has an octahedral shape and bond angles of $90°$ ✓. When Cl^- ions are added, a ligand substitution reaction takes place forming $CuCl_4^{2-}$ ✓ complex ions. $CuCl_4^{2-}$ has a tetrahedral shape and bond angles of $109.5°$. ✓

Ligand substitution involves one ligand being exchanged by another ligand: ✓

$$[Cu(H_2O)_6]^{2+} + 4Cl^- \rightleftharpoons CuCl_4^{2-} + 6H_2O \checkmark$$

In this type of response, there may be an additional mark for a clear, well-organised answer, using specialist terms. In addition, marks may be allocated for legible text with accurate spelling, punctuation and grammar.

Other types of questions

Free-response and open-ended questions allow you to choose the context and to develop your own ideas.

Multiple-choice or objective questions require you to select the correct response to the question from a number of given alternatives.

Stretch and Challenge

Stretch and Challenge is a concept that is applied to the structured questions in Units 4 and 5 of the exam papers for A2. In principle, it means that the sub-questions become progressively harder so as to challenge more able students and help differentiate between A and A* students.

Stretch and Challenge questions are designed to test a variety of different skills and your understanding of the material. They are likely to test your ability to make appropriate connections between different areas and apply your knowledge in unfamiliar contexts (as opposed to basic recall).

Exam technique

Links from AS

A2 Chemistry builds from the knowledge and understanding acquired after studying AS Chemistry. This Study Guide has been written so that you will be able to tackle A2 Chemistry from an AS Chemistry background.

You should not need to search for important chemistry from AS Chemistry because cross-references have been included as 'Key points from AS' to the AS Chemistry Study Guide: Revise AS.

What are examiners looking for?

Examiners use instructions to help you to decide the length and depth of your answer.

If a question does not seem to make sense, you may have misread it – read it again!

State, define or list

This requires a short, concise answer, often recall of material that can be learnt by rote.

Explain, describe or discuss

Some reasoning or some reference to theory is required, depending on the context.

Outline

This implies a short response, almost a list of sentences or bullet points.

Predict or deduce

You are not expected to answer by recall but by making a connection between pieces of information.

Suggest

You are expected to apply your general knowledge to a 'novel' situation, one which you have not directly studied during the A2 Chemistry course.

Calculate

This is used when a numerical answer is required. You should always use units in quantities and significant figures should be used with care.

Look to see how many significant figures have been used for quantities in the question and give your answer to this degree of accuracy.

If the question uses 3 sig figs, then give your answer to 3 sig figs also.

Exam technique

Some dos and don'ts

Dos

Do answer the question.
- No credit can be given for good Chemistry that is irrelevant to the question.

Do use the mark allocation to guide how much you write.
- Two marks are awarded for two valid points - writing more will rarely gain more credit and could mean wasted time or even contradicting earlier valid points.

Do use diagrams, equations and tables in your responses.
- Even in 'essay-type' questions, these offer an excellent way of communicating chemistry.

Do write legibly.
- An examiner cannot give marks if the answer cannot be read.

Do write using correct spelling and grammar. Structure longer essays carefully.
- Marks are now awarded for the quality of your language in exams.

Don'ts

Don't fill up any blank space on a paper.
- In structured questions, the number of dotted lines should guide the length of your answer.
- If you write too much, you waste time and may not finish the exam paper. You also risk contradicting yourself.

Don't write out the question again.
- This wastes time. The marks are for the answer!

Don't contradict yourself.
- The examiner cannot be expected to choose which answer is intended.

Don't spend too much time on a part that you find difficult.
- You may not have enough time to complete the exam. You can always return to a difficult calculation if you have time at the end of the exam.

What grade do you want?

Everyone would like to improve their grades but you will only manage this with a lot of hard work and determination. You should have a fair idea of your natural ability and likely grade in Chemistry and the hints below offer advice on improving that grade.

For a Grade A

You will need to be a very good all-rounder.

- You must go into every exam knowing the work extremely well.
- You must be able to apply your knowledge to new, unfamiliar situations.
- You need to have practised many, many exam questions so that you are ready for the type of question that will appear.

The exams test all areas of the specification and any weaknesses in your Chemistry will be found out. There must be no holes in your knowledge and understanding. For a Grade A, you must be competent in all areas.

For a Grade C

You must have a reasonable grasp of Chemistry but you may have weaknesses in several areas and you will be unsure of some of the reasons for the Chemistry.

- Many Grade C candidates are just as good at answering questions as Grade A students but holes and weaknesses often show up in just some topics.
- To improve, you will need to master your weaknesses and you must prepare thoroughly for the exam. You must become a better all-rounder.

For a Grade E

You cannot afford to miss the easy marks. Even if you find Chemistry difficult to understand and would be happy with a Grade E, there are plenty of questions in which you can gain marks.

- You must memorise all definitions.
- You must practise exam questions to give yourself confidence that you do know some Chemistry. In exams, answer the parts of questions that you know first. You must not waste time on the difficult parts. You can always go back to these later.
- The areas of Chemistry that you find most difficult are going to be hard to score on in exams. Even in the difficult questions, there are still marks to be gained. Show your working in calculations because credit is given for a sound method. You can always gain some marks if you get part of the way towards the solution.

What marks do you need?

The table below shows how your average mark is transferred into a grade.

average	80%	70%	60%	50%	40%
grade	A	B	C	D	E

The new A* grade

AQA have now introduced an A* grade for A level. This follows on from the introduction of this grade at GCSE and is intended to be awarded to a relatively small number of the highest scoring candidates.

To achieve an A* grade you must score over 80% on all six units combined (grade A) but also 90% on the three A2 units combined. This is quite a difficult target!

Four steps to successful revision

Step 1: Understand

- Study the topic to be learned slowly. Make sure you understand the logic or important concepts.
- Mark up the text if necessary – underline, highlight and make notes.
- Re-read each paragraph slowly.

GO TO STEP 2

Step 2: Summarise

- Now make your own revision note summary:
  What is the main idea, theme or concept to be learned?
  What are the main points? How does the logic develop?
  Ask questions: Why? How? What next?
- Use bullet points, mind maps, patterned notes.
- Link ideas with mnemonics, mind maps, crazy stories.
- Note the title and date of the revision notes
  (e.g. Chemistry: Reaction rates, 3rd March).
- Organise your notes carefully and keep them in a file.

This is now in **short-term memory**. You will forget 80% of it if you do not go to Step 3.
GO TO STEP 3, but first take a 10 minute break.

Step 3: Memorise

- Take 25 minute learning 'bites' with 5 minute breaks.
- After each 5 minute break test yourself:
  Cover the original revision note summary
  Write down the main points
  Speak out loud (record on tape)
  Tell someone else
  Repeat many times.

The material is well on its way to **long-term memory**.
You will forget 40% if you do not do step 4. **GO TO STEP 4**

Step 4: Track/Review

- Create a Revision Diary (one A4 page per day).
- Make a revision plan for the topic, e.g. 1 day later, 1 week later, 1 month later.
- Record your revision in your Revision Diary, e.g.
  Chemistry: Reaction rates, 3rd March 25 minutes
  Chemistry: Reaction rates, 5th March 15 minutes
  Chemistry: Reaction rates, 3rd April 15 minutes
  ... and then at monthly intervals.

Chapter 1
Rates and equilibria

The following topics are covered in this chapter:

- Orders and the rate equation
- Determination of reaction mechanisms
- The equilibrium constant, K_c
- Acids and bases
- The pH scale
- pH changes

1.1 Orders and the rate equation

After studying this section you should be able to:

- explain the terms 'rate of reaction', 'order', 'rate constant', 'rate equation'
- construct a rate equation and calculate its rate constant
- use the initial rates method to deduce orders
- understand that a temperature change increases the rate constant

Key points from AS

- Reaction rates
  Revise AS pages 83–86

During the study of reaction rates in AS Chemistry, reaction rates were described in terms of activation energy and the Boltzmann distribution. For A2 Chemistry, you will build upon this knowledge and understanding by measuring and calculating reaction rates using rate equations.

Orders and the rate equation

AQA M4

The rate of a reaction is determined from experimental results. Reaction rate **cannot** be determined from a chemical equation.

Suitable experimental techniques used to obtain rate data for a given reaction include:
- titration
- pH measurement
- colorimetry
- measuring volumes of gases evolved
- measuring mass changes.

The rate of a reaction is usually determined by the **change in concentration** of a reaction species **with time**.

- Units of rate = $\underbrace{\text{mol dm}^{-3}}_{\text{concentration}}\ \underbrace{\text{s}^{-1}}_{\text{per time}}$

Key points from AS

- What is a reaction rate?
  Revise AS pages 83–84

[A] means the concentration of reactant **A** in mol dm^{-3}.

Orders of reaction

For a reaction: **A** + **B** + **C** ⟶ products,
- the **order** of the reaction shows how the reaction rate is affected by the concentrations of **A**, **B** and **C**.

If the order is 0 (zero order) with respect to reactant **A**,
- the rate is unaffected by changes in concentration of **A**:
  $rate \propto [\mathbf{A}]^0$

If the order is 1 (first order) with respect to a reactant **B**,
- the rate is doubled by doubling of the concentration of reactant **B**:
  $rate \propto [\mathbf{B}]^1$

If the order is 2 (second order) with respect to a reactant **C**,
- the rate is quadrupled by doubling the concentration of reactant **C**:
  $rate \propto [\mathbf{C}]^2$

13

Rates and equilibria

> A number raised to the power of zero = 1 and zero order species can be omitted from the rate equation.

Combining this information,

$$rate \propto [A]^0[B]^1[C]^2$$
$$\therefore rate = k[B]^1[C]^2$$

- This expression is called the **rate equation** for the reaction.
- k is the rate constant of the reaction.

The rate equation

> The rate of a reaction is determined from experimental results.
>
> It **cannot** be determined from the chemical equation.

The rate equation of a reaction shows how the rate is affected by the concentration of reactants. A rate equation can only be determined from experiments.

In general, for a reaction: $A + B \longrightarrow C$,
the reaction rate is given by: $rate = k[A]^m[B]^n$

- m and n are the **orders of reaction** with respect to **A** and **B** respectively.
- The **overall order** of reaction is $m + n$.
- The reaction rate is measured as the change in concentration of a reaction species with time.
- The units of rate are $mol\ dm^{-3}\ s^{-1}$.

The rate constant k indicates the rate of the reaction:

- a large value of $k \longrightarrow$ fast rate of reaction
- a small value of $k \longrightarrow$ slow rate of reaction.

Determination of a rate of equation using the initial rates method

AQA M4

For a reaction, $X + Y \longrightarrow Z$, a series of experiments can be carried out using different **initial** concentrations of the reactants **X** and **Y**.

It is important to change only one variable at a time, so two series of experiments will be required.

- **Series 1** – The concentration of **X is changed** whilst the concentration of **Y** is kept constant.
- **Series 2** – The concentration of **Y is changed** whilst the concentration of **X** is kept constant.

> Clock reactions give an approximation to the initial rates method.

Clock reactions are often used to obtain a value for the initial rate of a reaction. A clock reaction measures the **time**, t, from the start of the reaction until a visual change – usually a change in colour, appearance of a precipitate or a solid disappearing.

- The clock reaction is repeated several times whilst varying the concentration of one of the reactants. The concentration of any other reactant is kept constant.
- The **initial rate** of the reaction is approximately proportional to $1/t$.

The results below show the initial rates using different concentrations of **X** and **Y**.

Experiment	[X(aq)] /mol dm^{-3}	[Y(aq)] /mol dm^{-3}	initial rate /mol dm^{-3} s^{-1}
1	1.0×10^{-2}	1.0×10^{-2}	0.5×10^{-3}
2	2.0×10^{-2}	1.0×10^{-2}	2.0×10^{-3}
3	2.0×10^{-2}	2.0×10^{-2}	4.0×10^{-3}

Determine the orders

Comparing experiments 1 and 2: [Y(aq)] has been kept constant

- [X(aq)] has been doubled,   rate × 4
  $\therefore$ order with respect to **X(aq)** = 2.

Rates and equilibria

Comparing experiments 2 and 3: [**X**(aq)] has been kept constant
- [**Y**(aq)] has been doubled, rate doubles
  ∴ order with respect to **Y**(aq) = 1.

Use the orders to write the rate equation
- rate = $k[\mathbf{X}(aq)]^2[\mathbf{Y}(aq)]$
- The overall order of this reaction is (2 + 1) = third order

Calculate the rate constant for the reaction

Rearrange the rate equation:

$$\text{The rate constant, } k = \frac{rate}{[\mathbf{X}(aq)]^2[\mathbf{Y}(aq)]}$$

Calculate k using values from one of the experiments.

$$k = \frac{(2.0 \times 10^{-3})}{(2.0 \times 10^{-2})^2 (1.0 \times 10^{-2})} = 500 \text{ dm}^6 \text{ mol}^{-2} \text{ s}^{-1}$$

> In this example, the results from Experiment 2 have been used.
> You will get the same value of k from the results of **any** of the experimental runs.
> Try calculating k from Experiment 1 and from Experiment 3. All give the same value.

Units of rate constants

The units of a rate constant depend upon the rate equation for the reaction. We can determine units of k by substituting units for rate and concentration into the rearranged rate equation. The table below shows how units can be determined.

> Notice how units need to be worked out afresh for reactions with different overall orders.
> You do not need to memorise these but it is important that you are able to work these out when needed.
> See also page 19 in which the units of the equilibrium constant K_c are discussed.

For a **zero-order** reaction:	For a **first-order** reaction:
rate = $k[A]^0 = k$ units of k = **mol dm^{-3} s^{-1}**	rate = $k[A]$ ∴ $k = \dfrac{rate}{[A]}$ Units of $k = \dfrac{(\text{mol dm}^{-3} \text{ s}^{-1})}{(\text{mol dm}^{-3})} = \mathbf{s^{-1}}$
For a **second-order** reaction:	For a **third-order** reaction:
rate = $k[A]^2$ ∴ $k = \dfrac{rate}{[A]^2}$ Units of $k = \dfrac{(\text{mol dm}^{-3} \text{ s}^{-1})}{(\text{mol dm}^{-3})^2}$ = **dm^3 mol^{-1} s^{-1}**	rate = $k[A]^2[B]$ ∴ $k = \dfrac{rate}{[A]^2[B]}$ Units of $k = \dfrac{(\text{mol dm}^{-3} \text{ s}^{-1})}{(\text{mol dm}^{-3})^2 (\text{mol dm}^{-3})}$ = **dm^6 mol^{-2} s^{-1}**

How does the rate constant, k, vary with temperature?

AQA M4

During AS Chemistry, you studied how the rate of reaction increases with temperature in terms of collisions, the activation energy of the reaction and the Boltzmann distribution of molecular energies.

> **Key points from AS**
> - Boltzmann distribution
> - Activation energy
> *Revise AS page 84*

The effect of temperature on rate constants

An increase in temperature speeds up the rate of most reactions and the rate constant, k, increases.

The table below shows the increase in the rate constant for the decomposition of hydrogen iodide with increasing temperature.

$$2HI(g) \longrightarrow H_2(g) + I_2(g)$$

> For many reactions, the rate doubles for each 10°C increase in temperature.

temperature/K	556	629	700	781
rate constant, k /dm^3 mol^{-1} s^{-1}	7.04×10^{-7}	6.04×10^{-5}	2.32×10^{-3}	7.90×10^{-2}

Rates and equilibria

A reaction with a large activation energy has a small rate constant. Such a reaction may need a considerable temperature rise to increase the value of the rate constant sufficiently for a reaction to take place at all.

Why are catalysts so effective?

A catalyst is so effective because:

- it reduces the **activation energy**
- which increases the **rate of reaction**
- which increases the **rate constant, k**.

Progress check

1. A chemical reaction is first order with respect to compound **P** and second order with respect to compound **Q**.
   (a) Write the rate equation for this reaction.
   (b) What is the overall order of this reaction?
   (c) By what factor will the rate increase if:
      (i) the concentration of **P** only is doubled
      (ii) the concentration of **Q** only is doubled
      (iii) the concentrations of **P** and **Q** are **both** doubled?
   (d) What are the units of the rate constant of this reaction?

2. The reaction of ethanoic anhydride, $(CH_3CO)_2O$, with ethanol, C_2H_5OH, can be represented by the equation:

$$(CH_3CO)_2O + C_2H_5OH \longrightarrow CH_3CO_2C_2H_5 + CH_3CO_2H$$

The table below shows the initial concentrations of the two reactants and the initial rates of reaction.

Experiment	$[(CH_3CO)_2O]$ /mol dm^{-3}	$[C_2H_5OH]$ /mol dm^{-3}	initial rate /mol dm^{-3} s^{-1}
1	0.400	0.200	6.60×10^{-4}
2	0.400	0.400	1.32×10^{-3}
3	0.800	0.400	2.64×10^{-3}

(a) State and explain the order of reaction with respect to:
   (i) ethanoic anhydride (ii) ethanol.
(b) (i) Write an expression for the overall rate equation.
    (ii) What is the overall order of reaction?
(c) Calculate the value, with units, for the rate constant, k.

1 (a) rate = $k[P][Q]^2$
  (b) third order
  (c) (i) rate doubles (ii) rate quadruples (iii) rate × 8
  (d) dm^6 mol^{-2} s^{-1}
2 (a) (i) first order; concentration doubles; rate doubles
     (ii) first order; concentration doubles; rate doubles.
  (b) (i) rate = k [(CH$_3$CO)$_2$O] [C$_2$H$_5$OH] (ii) second order.
  (c) 8.25×10^{-3} dm^3 mol^{-1} s^{-1}.

Rates and equilibria

1.2 Determination of reaction mechanisms

After studying this section you should be able to:

- understand what is meant by a rate-determining step
- predict a rate equation from a rate-determining step
- predict a rate-determining step from a rate equation

Predicting reaction mechanisms

AQA   M4

Key points from AS

- **The hydrolysis of halogenoalkanes**
  Revise AS pages 115–116

You are **not** expected to remember these examples.

You **are** expected to interpret data to identify a rate-determining step and to suggest possible steps in the mechanism for the reaction.

This reaction proceeds via a two-step mechanism:
1 particle,
$(CH_3)_3CBr$,
is involved in the rate-determining step of a Nucleophilic Substitution mechanism.

The rate-determining step

Chemical reactions often take place in a series of steps. The detail of these steps is the **reaction mechanism**.

> The rate equation can provide clues about a likely reaction mechanism by identifying the **slowest** stage of a reaction sequence, called the **rate-determining step**.
>
> **KEY POINT**

The hydrolysis of tertiary halogenoalkanes

2-bromo-2-methylpropane, $(CH_3)_3C–Br$, is hydrolysed by aqueous alkali, OH^-.

Experiments show that the rate equation for this reaction is:

$$rate = k[(CH_3)_3C–Br]$$

- The rate equation shows that a **slow** reaction step must involve **only** $(CH_3)_3C–Br$.
- OH^- has **no** effect on the reaction rate.

This supports the slow step below:

$$\underbrace{(CH_3)_3C–Br}_{\text{One particle in rate-determining step}} \longrightarrow (CH_3)_3C^+ + Br^- \qquad \textbf{SLOW}$$

The overall equation for the hydrolysis of 1-bromobutane by aqueous alkali is shown below:

$$(CH_3)_3C–Br + OH^- \longrightarrow (CH_3)_3C–OH + Br^-$$

- This **does not** match the equation for the slow step above.
- The rate-determining step must be followed by further **fast** steps.
- OH^- must be involved as it is in the **overall equation**.

A possible second step is shown below:

$$(CH_3)_3C^+ + OH^- \longrightarrow (CH_3)_3C–OH \qquad \textbf{FAST}$$

The **overall equation** is the sum of the equations from each step in the mechanism.

Notice that $(CH_3)_3C^+$ does not feature in the overall equation. It is generated in one step and used up in another step.

In the hydrolysis of a tertiary halogenoalkane:

1st step	$(CH_3)_3C–Br$	$\longrightarrow (CH_3)_3C^+ + Br^-$	**SLOW**
2nd step	$(CH_3)_3C^+ + OH^-$	$\longrightarrow (CH_3)_3C–OH$	**FAST**
Overall equation	$(CH_3)_3C–Br + OH^-$	$\longrightarrow (CH_3)_3C–OH + Br^-$	

The hydrolysis of primary halogenoalkanes

1-Bromobutane, $CH_3CH_2CH_2CH_2Br$, is hydrolysed by aqueous alkali, OH^-.

Experiments show that the rate equation for this reaction is:

$$rate = k[CH_3CH_2CH_2CH_2Br][OH^-]$$

Rates and equilibria

> This reaction proceeds via a one-step mechanism:
> 2 particles, $CH_3CH_2CH_2CH_2Br$ and OH^-, are involved in the rate-determining step of a nucleophilic Substitution mechanism.

The **slow** reaction step must involve **both** $CH_3CH_2CH_2CH_2Br$ and OH^-.
This supports the slow step below:

$$\underbrace{CH_3CH_2CH_2CH_2Br + OH^- \longrightarrow CH_3CH_2CH_2CH_2OH + Br^-}_{\text{Two particles in rate-determining step}} \quad \textbf{SLOW}$$

The overall equation for the hydrolysis of 1-bromobutane by aqueous alkali is:

$$CH_3CH_2CH_2CH_2Br + OH^- \longrightarrow CH_3CH_2CH_2CH_2OH + Br^-$$

- This **does** match the equation for the slow step above.
- The reaction **mechanism** must have just a **single step**.

The reaction of iodine and propanone in acid

The reaction of iodine, I_2, and propanone, CH_3COCH_3, is catalysed by acid, H^+.
The overall equation for the acid-catalysed reaction of iodine and propanone is:

$$I_2 + CH_3COCH_3 \longrightarrow CH_3COCH_2I + H^+ + I^- \quad \text{overall equation}$$

Experiments show that the rate equation for this reaction is:

$$rate = k[CH_3COCH_3][H^+]$$

> The slow step is the **rate-determining step**.

The **slow** rate-determining step must involve **both** CH_3COCH_3 and H^+.
A possible equation for the rate-determining step is shown below:

$$\underbrace{CH_3COCH_3 + H^+ \longrightarrow CH_3C(OH^+)CH_3}_{\text{Two particles in rate-determining step}} \quad \textbf{SLOW}$$

> If the species in the rate equation do not match those in the overall equation, the reaction mechanism must have more than one step.

- This does **not** match the overall equation.

Possible further steps are shown below:

$$CH_3C(OH^+)CH_3 \longrightarrow CH_2=C(OH)CH_3 + H^+ \quad \textbf{FAST}$$
$$CH_3C(OH)=CH_2 + I_2 \longrightarrow CH_3(OH)CICH_2I \quad \textbf{FAST}$$
$$CH_3(OH)CICH_2I \longrightarrow CH_3COCH_2I + H^+ + I^- \quad \textbf{FAST}$$

> H^+ is a catalyst.
> It is needed for the first step but is regenerated in a further step.
> Overall, it is not consumed.

> **KEY POINT**
> The overall equation is the sum of the equations from each step in the mechanism.

- The rate of the reaction is controlled mainly by the slowest step of the mechanism – the **rate-determining step**.
- The rate equation only includes reacting species involved in the slow rate-determining step.
- The orders in the rate equation match the number of species involved in the rate-determining step.

Progress check

1. CH_3CH_2Br and cyanide ions, CN^- react as follows:
   $$CH_3CH_2Br + CN^- \longrightarrow CH_3CH_2CN + Br^-$$
   Show two different possible reaction routes for this reaction and write down the expected rate equation for each route.

   1 Single step: $CH_3CH_2Br + CN^- \longrightarrow CH_3CH_2CN + Br^-$ rate = $k[CH_3CH_2Br][CN^-]$
   Two steps: $CH_3CH_2Br \longrightarrow CH_3CH_2^+ + Br^-$ slow
   $CH_3CH_2^+ + CN^- \longrightarrow CH_3CH_2CN$ fast rate = $k[CH_3CH_2Br]$

Rates and equilibria

1.3 The equilibrium constant, K_c

After studying this section you should be able to:

- deduce, for homogeneous reactions, K_c in terms of concentrations
- calculate values of K_c given appropriate data
- understand that K_c is changed only by changes in temperature
- understand how the magnitude of K_c relates to the equilibrium position
- calculate, from data, the concentrations present at equilibrium

Key points from AS

- Chemical equilibrium
  Revise AS pages 88–92

During the study of equilibrium in AS Chemistry, the concept of dynamic equilibrium is introduced and le Chatelier's principle is used to predict how a change in conditions may alter the equilibrium position. For A2 Chemistry, you will build upon this knowledge and understanding to find out the exact position of equilibrium using the Equilibrium Law.

The equilibrium law

AQA M4

The equilibrium law states that, for an equation:

$$a\mathbf{A} + b\mathbf{B} \rightleftharpoons c\mathbf{C} + d\mathbf{D},$$

$$K_c = \frac{[\mathbf{C}]^c [\mathbf{D}]^d}{[\mathbf{A}]^a [\mathbf{B}]^b}$$

- $[\mathbf{A}]^a$, etc., are the **equilibrium** concentrations of the reactants and products of the reaction.
- Each product and reactant has its equilibrium concentration raised to the **power** of its **balancing number** in the equation.
- The equilibrium concentrations of the **products** are multiplied together on **top** of the fraction.
- The equilibrium concentrations of the **reactants** are multiplied together on the **bottom** of the fraction.

The overall equation for the reaction is used to determine K_c.

Working out K_c

For the equilibrium: $H_2(g) + I_2(g) \rightleftharpoons 2HI(g)$

applying the equilibrium law above: $K_c = \dfrac{[HI(g)]^2}{[H_2(g)][I_2(g)]}$

*In exams, you are likely to be assessed only on **homogeneous equilibria** – all species are in the same phase:*

all gaseous (g)
or
all aqueous (aq)
or
all liquid (l).

Equilibrium concentrations of $H_2(g)$, $I_2(g)$ and $HI(g)$ are shown below:

$[H_2(g)]$ /mol dm^{-3}	$[I_2(g)]$ /mol dm^{-3}	$[HI(g)]$ /mol dm^{-3}
0.0114	0.00120	0.0252

$$K_c = \frac{[HI(g)]^2}{[H_2(g)][I_2(g)]} = \frac{0.0252^2}{0.0114 \times 0.00120} = 46.4$$

Units of K_c

The units must be worked out afresh for each equilibrium. See also page 15 in which the units of the rate constant are discussed.

- In the K_c expression, each concentration value is replaced by its units:

$$K_c = \frac{[HI(g)]^2}{[H_2(g)][I_2(g)]} = \frac{(\text{mol dm}^{-3})^2}{(\text{mol dm}^{-3})(\text{mol dm}^{-3})}$$

- For this equilibrium, the units cancel.

$\therefore$ in the equilibrium: $H_2(g) + I_2(g) \rightleftharpoons 2HI(g)$, K_c has no units.

Rates and equilibria

Properties of K_c

The magnitude of K_c indicates the extent of a chemical equilibrium.

- $K_c = 1$ indicates an equilibrium halfway between reactants and products.
- $K_c = 100$ indicates an equilibrium well in favour of the products.
- $K_c = 1 \times 10^{-2}$ indicates an equilibrium well in favour of the reactants.

K_c indicates how FAR a reaction proceeds; **not** how FAST.
K_c is unaffected by changes in concentration or pressure.
K_c can **only** be changed by altering the temperature.

> Although changes in K_c affect the equilibrium yield, the actual conditions used may be different.
>
> If the forward reaction is **exothermic**, K_c increases and the equilibrium yield increases as temperature decreases.
>
> However, a decrease in temperature may slow down the reaction so much that the reaction is stopped.
>
> In practice, a compromise will be needed where equilibrium and rate are considered together – a reasonable equilibrium yield must be obtained in a reasonable length of time.

How does K_c vary with temperature?

It cannot be stressed too strongly that K_c can be changed only by altering the temperature. How K_c changes depends upon whether the reaction gives out or takes in heat energy.

If the forward reaction is **exothermic**, K_c **decreases** with increasing temperature. Raising the temperature reduces the equilibrium yield of products.

$$H_2(g) + I_2(g) \rightleftharpoons 2HI(g): \qquad \Delta H^\ominus_{298} = -9.6 \text{ kJ mol}^{-1}$$

temperature /K	500	700	1100
K_c	160	54	25

If the forward reaction is **endothermic**, K_c **increases** with increasing temperature. Raising the temperature increases the equilibrium yield of products.

$$N_2(g) + O_2(g) \rightleftharpoons 2NO(g): \qquad \Delta H^\ominus_{298} = +180 \text{ kJ mol}^{-1}$$

temperature /K	500	700	1100
K_c	5×10^{-13}	4×10^{-8}	1×10^{-5}

Progress check

1. For each of the following equilibria, write down the expression for K_c. State the units for K_c for each reaction.
   (a) $N_2O_4(g) \rightleftharpoons 2NO_2(g)$
   (b) $CO(g) + 2H_2(g) \rightleftharpoons CH_3OH(g)$
   (c) $H_2(g) + Br_2(g) \rightleftharpoons 2HBr(g)$
   (d) $2SO_2(g) + O_2(g) \rightleftharpoons 2SO_3(g)$

2. Explain whether the two reactions, **A** and **B**, are exothermic or endothermic.

temperature /K	numerical value of K_c	
	reaction A	reaction B
200	5.51×10^{-8}	4.39×10^4
400	1.46	4.03
600	3.62×10^2	3.00×10^{-2}

1. (a) $K_c = \dfrac{[NO_2(g)]^2}{[N_2O_4(g)]}$ mol dm^{-3}
   (b) $K_c = \dfrac{[CH_3OH(g)]}{[CO(g)][H_2(g)]^2}$ dm^6 mol^{-2}
   (c) $K_c = \dfrac{[H_2(g)][Br_2(g)]}{[HBr(g)]^2}$
   (d) $K_c = \dfrac{[SO_3(g)]^2}{[SO_2(g)]^2[O_2(g)]}$ dm^3 mol^{-1}

2. Reaction A: endothermic; K_c increases with increasing temperature.
   Reaction B: exothermic; K_c decreases with increasing temperature.

Rates and equilibria

Determination of K_c from experiment

AQA M4

The equilibrium constant K_c can be calculated using experimental results. The example below shows how to:

- determine the equilibrium concentrations of the components in an equilibrium mixture
- calculate K_c.

The ethyl ethanoate esterification equilibrium

*The most important stage in this calculation is to find the **change** in the number of moles of each species in the equilibrium.*

0.200 mol CH_3COOH and 0.100 mol C_2H_5OH were mixed together with a trace of acid catalyst in a total volume of 250 cm^3. The mixture was allowed to reach equilibrium:

$$CH_3COOH + C_2H_5OH \rightleftharpoons CH_3COOC_2H_5 + H_2O.$$

Analysis of the mixture showed that 0.115 mol of CH_3COOH were present at equilibrium.

First summarise the results

It is useful to summarise the results as an 'I.C.E.' table: 'Initial/Change/Equilibrium'.

	CH_3COOH	C_2H_5OH	$CH_3COOC_2H_5$	H_2O
Initial no. of moles	0.200	0.100	0	0
Change in moles				
Equilibrium no. of moles	0.115			

From the CH_3COOH values:

- moles of CH_3COOH that reacted = 0.200 − 0.115 = **0.085** mol

Find the change in moles of each component in the equilibrium

The amount of each component can be determined from the balanced equation. The change in moles of CH_3COOH is already known.

*In this experiment, 0.085 mol CH_3COOH has **reacted** with 0.085 mol C_2H_5OH to form 0.085 mol $CH_3COOC_2H_5$ and 0.085 mol H_2O.*

equation:	CH_3COOH	+	C_2H_5OH	$\rightleftharpoons$	$CH_3COOC_2H_5$	+	H_2O
molar quantities:	1 mol		1 mol	$\longrightarrow$	1 mol		1 mol
change/mol:	−0.085		−0.085		+0.085		+0.085

Equilibrium concentrations are determined for each component

Note that the total volume is 250 cm^3 (0.250 dm^3). The concentration must be expressed as mol dm^{-3}.

	CH_3COOH +	C_2H_5OH $\rightleftharpoons$	$CH_3COOC_2H_5$ +	H_2O
Initial amount/mol	0.200	0.100	0	0
Change in moles	−0.085	−0.085	+0.085	+0.085
Equilibrium amount/mol	0.115	0.015	0.085	0.085
Equilibrium conc. /mol dm^{-3}	$\dfrac{0.115}{0.250}$	$\dfrac{0.015}{0.250}$	$\dfrac{0.085}{0.250}$	$\dfrac{0.085}{0.250}$

Write the expression for K_c and substitute values

$$K_c = \frac{[CH_3COOC_2H_5][H_2O]}{[CH_3COOH][C_2H_5OH]} = \frac{\dfrac{0.085}{0.250} \times \dfrac{0.085}{0.250}}{\dfrac{0.115}{0.250} \times \dfrac{0.015}{0.250}}$$

Calculate K_c

$\therefore K_c = 4.19$ (no units: all units cancel)

Rates and equilibria

Progress check

1. Several experiments were set up for the $H_2(g)$, $I_2(g)$ and $HI(g)$ equilibrium. The equilibrium concentrations are shown below.

$[H_2(g)]$ /mol dm^{-3}	$[I_2(g)]$ /mol dm^{-3}	$[HI(g)]$ /mol dm^{-3}
0.0092	0.0020	0.0296
0.0077	0.0031	0.0334
0.0092	0.0022	0.0308
0.0035	0.0035	0.0235

   Calculate the value for K_c for each experiment. Hence, show that each experiment has the same value for K_c (allowing for experimental error). Work out an average value for K_c.

2. 2 moles of ethanoic acid, CH_3COOH were mixed with 3 moles of ethanol and the mixture was allowed to reach equilibrium.
$$CH_3COOH + C_2H_5OH \rightleftharpoons CH_3COOC_2H_5 + H_2O$$
   At equilibrium, 0.5 moles of ethanoic acid remained.
   (a) Work out the equilibrium concentrations of each component in the mixture. (You will need to use V to represent the volume but this will cancel out in your calculation.)
   (b) Use these values to calculate K_c.

3. When 0.50 moles of $H_2(g)$ and 0.18 moles of $I_2(g)$ were heated at 500°C, the equilibrium mixture contained 0.01 moles of $I_2(g)$.
   The equation is:
   $H_2(g) + I_2(g) \rightleftharpoons 2HI(g)$
   (a) How many moles of $I_2(g)$ reacted?
   (b) How many moles of $H_2(g)$ were present at equilibrium?
   (c) How many moles of $HI(g)$ were present at equilibrium?
   (d) Calculate the equilibrium constant, K_c for this reaction.

1 values: 47.6; 46.7; 46.9; 45.1. average = 46.6
2 (a) CH_3COOH, 0.5/V mol dm^{-3}; C_2H_5OH, 1.5/V mol dm^{-3}; $CH_3COOC_2H_5$, 1.5/V mol dm^{-3}; H_2O, 1.5/V mol dm^{-3}.
(b) K_c = 3.
3 (a) 0.17 mol (b) 0.33 mol (c) 0.34 mol (d) 35.

Applying rates and equilibrium to industrial processes

Industrial processes need to obtain economic yields of the product being manufactured. The balance between **enthalpy**, **entropy** and **temperature** is important.

Equilibrium

> You do not need to know the actual industrial conditions used in the Haber process.

In the Haber process for the production of ammonia

$$N_2(g) + 3H_2(g) \rightleftharpoons 2NH_3(g) \qquad \Delta H^\ominus = -92 \text{ kJ mol}^{-1}$$

- the right-hand side has **fewer moles of gas** than the left-hand side (reactants), favoured by a **high** pressure
- the forward reaction is **exothermic**, favoured by a **low** temperature.

> We can arrive at these conclusions by using le Chatelier's principle from AS Chemistry.

The optimum equilibrium conditions for maximum equilibrium yield are: **high pressure and low temperature**.

Rates and equilibria

> The rate of a reaction increases with higher temperatures and pressures – but temperature has a much larger effect.

The need for compromise

In reality it may not be feasible to use optimum equilibrium conditions.

- Compressing gases to **high pressures** has a high **energy cost**. There are also considerable **safety** implications of using very high pressures.
- At **low temperatures**, the rate of reaction is **slow** as few molecules possess the necessary activation energy of the reaction.

There needs to be a compromise.

> **Compromising**
> Optimum conditions, reaction rate, feasibility, reality, safety and economics must be considered.

- Use of a **high enough pressure** to **increase** the equilibrium yield without increasing **energy costs** in generating the pressure or compromising **safety**.
- Increase the temperature just enough to increase the rate of reaction so that the reaction takes place in a realistic time frame, whilst still producing an **adequate equilibrium yield**.

Equilibrium is never reached

Another factor is that industrial processes do not operate at equilibrium:

- products are removed – we never have a closed system
- there is insufficient time for equilibrium to be reached.

But it is still important to consider the conditions needed to secure a high equilibrium yield.

The search for better processes

The chemical industry takes steps to maximise the **atom economy** of the process. Many factors need to be considered, including:

- **recycling** unreacted reagents
- developing **alternative reactions** and processes
- using **more available** raw materials, preferably **renewable**
- using **catalysts** to allow processes to occur at **lower temperatures**, cutting energy needs
- developing **simpler processes** requiring less costly chemical plants
- using **co-products** and **by-products**, rather than disposing of them.

> Hazards can be reduced by developing processes that reduce risks from toxic reactants, explosions, acidic or flammable gases, toxic emissions, etc.

Rates and equilibria

1.4 Acids and bases

After studying this section you should be able to:

- describe what is meant by Brønsted–Lowry acids and bases
- understand conjugate acid–base pairs
- understand the difference between a strong and a weak acid
- define the acid dissociation constant, K_a

Key points from AS

- Calculations in acid–base titrations
  Revise AS page 32

During the study of acids and bases at GCSE, you learnt that the pH scale can be used to measure the strength of acids and bases. You also learnt the reactions of acids with metals, carbonates and alkalis. During AS Chemistry, you revisited these reactions and found out how titrations can be used to measure the concentration of an unknown acid or base. For A2 Chemistry, you will study acids in terms of proton transfer, strength, pH and buffers.

Brønsted–Lowry acids and bases

A molecule of an acid contains a hydrogen atom that can be released as a positive hydrogen ion or proton, H^+.

- An acid is a substance producing an excess of hydrogen ions in solution – the Arrhenius theory.
- An acid is a proton donor – the Brønsted–Lowry theory.

At A level, we use the Brønsted–Lowry model of acids and bases in terms of proton transfer.

An acid is a proton donor.

A base is a proton acceptor.

- A Brønsted–Lowry **acid** is a proton donor.
- A Brønsted–Lowry **base** is a proton acceptor.
- An **alkali** is a base that dissolves in water forming $OH^-(aq)$ ions.

Acid–base pairs

Acids and bases are linked by H^+ as **conjugate pairs**:

- the **conjugate acid** donates H^+
- the **conjugate base** accepts H^+.

$$HCl \rightleftharpoons H^+ + Cl^-$$
conjugate acid → conjugate base

Examples of some conjugate acid–base pairs are shown below.

	acid		base
hydrochloric acid	HCl	$\rightleftharpoons H^+ +$	Cl^-
sulfuric acid	H_2SO_4	$\rightleftharpoons H^+ +$	HSO_4^-
ethanoic acid	CH_3COOH	$\rightleftharpoons H^+ +$	CH_3COO^-

Rates and equilibria

An acid needs a base

An acid can only donate a proton if there is a base to accept it. Most reactions of acids take place in aqueous conditions with water acting as the base. By mixing an acid with a base, an equilibrium is set up comprising **two** acid–base **conjugate pairs**.

The equilibrium system in aqueous ethanoic acid is shown below.

You should be able to identify acid–base pairs in equations such as this.

- In a reaction involving an aqueous acid such as hydrochloric acid, the 'active' species is the **oxonium** ion, H_3O^+ (ACID 2 above).
- Formation of the oxonium ion requires **both** an acid **and** water.

In equations, the oxonium ion H_3O^+ is usually shown as $H^+(aq)$.

Progress check

1. Identify the acid–base pairs in the acid–base equilibria below.
   (a) $HNO_3 + H_2O \rightleftharpoons H_3O^+ + NO_3^-$
   (b) $NH_3 + H_2O \rightleftharpoons NH_4^+ + OH^-$
   (c) $H_2SO_4 + H_2O \rightleftharpoons HSO_4^- + H_3O^+$

2. Write equations for the following acid–base equilibria:
   (a) hydrochloric acid and hydroxide ions
   (b) ethanoic acid, CH_3COOH, and water.

1 (a) acid 1: HNO_3, base 1: NO_3^-; acid 2: H_3O^+, base 2: H_2O
  (b) acid 1: H_2O, base 1: OH^-; acid 2: NH_4^+, base 2: NH_3
  (c) acid 1: H_2SO_4, base 1: HSO_4^-; acid 2: H_3O^+, base 2: H_2O
2 (a) $HCl + OH^- \rightleftharpoons H_2O + Cl^-$
  (b) $CH_3COOH + H_2O \rightleftharpoons H_3O^+ + CH_3COO^-$

Strength of acids and bases

AQA M4

The acid–base equilibrium of an acid, HA, in water is shown below.

$$HA(aq) + H_2O(l) \rightleftharpoons H_3O^+(aq) + A^-(aq)$$

To emphasise the loss of H^+, this can be shown more simply as **dissociation** of the acid HA:

$$HA(aq) \rightleftharpoons H^+(aq) + A^-(aq).$$

The **strength** of an acid is the extent that the acid dissociates into H^+ and A^-.

Strong acids

A strong acid is completely dissociated.

A **strong** acid, such as nitric acid, HNO_3, is a **good** proton donor.
- There is almost **complete** dissociation.

$$HNO_3(aq) \xrightleftharpoons[]{\text{equilibrium}} H^+(aq) + NO_3^-(aq)$$

A weak acid only partially dissociates.

- Virtually all of the potential acidic power has been released as $H^+(aq)$.
- At equilibrium, $[H^+(aq)]$ is much greater than $[HNO_3(aq)]$.

Rates and equilibria

Weak acids

A **weak** acid, such as ethanoic acid, CH_3COOH, is a **poor** proton donor.
- There is only **partial** dissociation.

$$CH_3COOH(aq) \underset{\text{equilibrium}}{\rightleftharpoons} H^+(aq) + CH_3COO^-(aq)$$

Only a small proportion of the potential acidic power has been released as $H^+(aq)$. At equilibrium, $[CH_3COOH(aq)]$ is much greater than $[H^+(aq)]$.

The acid dissociation constant, K_a

> K_a is just a special equilibrium constant K_c for equilibria showing the dissociation of acids.

The extent of acid dissociation is shown by the **acid dissociation constant**, K_a.
For the reaction: $HA(aq) \rightleftharpoons H^+(aq) + A^-(aq)$,

$$K_a = \frac{[H^+(aq)][A^-(aq)]}{[HA(aq)]}$$

> $[HA(aq)]$, $[H^+(aq)]$ and $[A^-(aq)]$ are equilibrium concentrations.

units: $K_a = \dfrac{(\text{mol dm}^{-3})^2}{(\text{mol dm}^{-3})} = \text{mol dm}^{-3}$.

> K_a is sometimes called the acidity constant.

- A large K_a value shows that the extent of **dissociation is large** – the acid is strong.
- A small K_a value shows that the extent of **dissociation is small** – the acid is weak.

Acid strength and concentration

> *Concentrated* and *dilute* are terms used to describe the **amount** of dissolved acid in a solution.
>
> *Strong* and *weak* are terms used to describe the degree of **dissociation** of an acid.

The distinction between the strength and concentration of an acid is important.

> **KEY POINT**
>
> **Concentration** is the **amount** of an acid dissolved in 1 dm^3 of solution.
> - Concentration is measured in mol dm^{-3}.
>
> **Strength** is the extent of **dissociation** of an acid.
> - Strength is measured as K_a in units determined from the equilibrium.

Progress check

1. For each of the following acid–base equilibria, write down the expression for K_a. State the units of K_a for each reaction.
   (a) $HCOOH(aq) \rightleftharpoons H^+(aq) + HCOO^-(aq)$
   (b) $C_6H_5COOH(aq) \rightleftharpoons H^+(aq) + C_6H_5COO^-(aq)$

2. Samples of two acids, hydrochloric acid and ethanoic acid, have the same concentration: 0.1 mol dm^{-3}. Explain why one 'dilute acid' is *strong* whereas the other 'dilute acid' is *weak*.

1. (a) $K_a = \dfrac{[H^+(aq)][HCOO^-(aq)]}{[HCOOH(aq)]}$ units: mol dm^{-3}
   (b) $K_a = \dfrac{[H^+(aq)][C_6H_5COO^-(aq)]}{[C_6H_5COOH(aq)]}$ units: mol dm^{-3}

2. Concentration applies to the amount, in mol, in 1 dm^3 of solution. Both solutions have 0.1 mol dissolved in 1 dm^3 of solution and are dilute. Hydrochloric acid is strong because its dissociation is near to complete. However, ethanoic acid only donates a small proportion of its potential protons, its dissociation is incomplete and it is a weak acid.

Rates and equilibria

1.5 The pH scale

After studying this section you should be able to:

- define the terms pH, pK_a and K_w
- calculate pH from $[H^+(aq)]$
- calculate $[H^+(aq)]$ from pH
- understand the meaning of the ionisation product of water, K_w
- calculate pH for strong bases

pH and $[H^+(aq)]$

The concentrations of $H^+(aq)$ ions in aqueous solutions vary widely between about 10 mol dm^{-3} and about 1×10^{-15} mol dm^{-3}.

The **pH scale** is a logarithmic scale used to overcome the problem of using this large range of numbers and to ease the use of negative powers.

The pH scale

pH	$[H^+]$ / mol dm^{-3}
0	1
1	1×10^{-1}
2	1×10^{-2}
3	1×10^{-3}
4	1×10^{-4}
5	1×10^{-5}
6	1×10^{-6}
7	1×10^{-7}
8	1×10^{-8}
9	1×10^{-9}
10	1×10^{-10}
11	1×10^{-11}
12	1×10^{-12}
13	1×10^{-13}
14	1×10^{-14}

KEY POINT

pH is defined as: $pH = -\log_{10} [H^+(aq)]$
- where $[H^+(aq)]$ is the concentration of hydrogen ions in aqueous solution.
- $[H^+(aq)]$ can be calculated from pH using:
$[H^+(aq)] = 10^{-pH}$

Notice how the value of pH is linked to the power of 10.

pH	2	9	3.6	10.3
$[H^+(aq)]$/mol dm^{-3}	10^{-2}	10^{-9}	$10^{-3.6}$	$10^{-10.3}$

What does a pH value mean?

- A **low** value of $[H^+(aq)]$ matches a **high** value of pH.
- A **high** value of $[H^+(aq)]$ matches a **low** value of pH.
- A change of pH by 1 changes $[H^+(aq)]$ by 10 times.
- An acid of pH 4 contains 10 times the concentration of $H^+(aq)$ ions as an acid of pH 5.

Calculating the pH of strong acids

For a strong acid, HA:
- we can assume complete dissociation
- the concentration of $H^+(aq)$ can be found directly from the acid concentration: $[H^+] = [HA]$.

HA is **monoprotic**: each HA molecule can donate **one** H^+ ion.

You should be able to convert pH into $[H^+(aq)]$ and *vice versa*.

Example 1
A strong acid, HA, has a concentration of 0.010 mol dm^{-3}. What is the pH?
Complete dissociation. $\therefore [H^+(aq)] = 0.010$ mol dm^{-3}
$$pH = -\log_{10} [H^+(aq)] = -\log_{10} (0.010) = \mathbf{2.0}$$

This is the key to use on your calculator when doing pH and $[H^+]$ calculations.
For 10^x, press the **SHIFT** or **INV** key first.

Example 2
A strong acid, HA, has a pH of 3.4. What is the concentration of $H^+(aq)$?
Complete dissociation. $\therefore [H^+(aq)] = 10^{-pH} = 10^{-3.4}$ mol dm^{-3}
$$\therefore [H^+(aq)] = \mathbf{3.98 \times 10^{-4}} \text{ mol dm}^{-3}$$

Rates and equilibria

Hints for pH calculations

Calculations involving pH are easy once you have learnt how to use your calculator properly.

- Try the examples on the previous page until you can remember the order to press the keys.
- Try reversing each calculation to go back to the original value. Repeat several times until you have mastered how to use **your** calculator for pH calculations.
- **Don't** borrow a calculator or you will get confused. Different calculators may need the keys to be pressed in a different order!
- Look at your answer and decide whether it looks sensible.

> **Learn:** $pH = -\log_{10}[H^+(aq)]$
> $[H^+(aq)] = 10^{-pH}$.

KEY POINT

Progress check

1. Calculate the pH of solutions with the following $[H^+(aq)]$ values.
   (a) 0.01 mol dm^{-3}
   (b) 0.0001 mol dm^{-3}
   (c) 1.0×10^{-13} mol dm^{-3}
   (d) 2.50×10^{-3} mol dm^{-3}
   (e) 8.10×10^{-6} mol dm^{-3}
   (f) 4.42×10^{-11} mol dm^{-3}

2. Calculate $[H^+(aq)]$ of solutions with the following pH values.
   (a) pH 3
   (b) pH 10
   (c) pH 2.8
   (d) pH 7.9
   (e) pH 12.2
   (f) pH 9.6

3. How many times more hydrogen ions are in an acid of pH 1 than an acid of pH 5?

4. How can a solution have a pH with a negative value?

4 A solution with $[H^+(aq)] > 1$ mol dm^{-3} has a negative pH value.
3 pH 1 has 10 000 times more H^+ ions than pH 5.
2 (a) 1×10^{-3} mol dm^{-3} (b) 1×10^{-10} mol dm^{-3} (c) 1.58×10^{-3} mol dm^{-3} (d) 1.26×10^{-8} mol dm^{-3} (e) 6.31×10^{-13} mol dm^{-3} (f) 2.51×10^{-10} mol dm^{-3}
1 (a) 2 (b) 4 (c) 13 (d) 2.60 (e) 5.09 (f) 10.4

Calculating the pH of weak acids

AQA M4

The pH of a weak acid HA can be calculated from:
- the **concentration** of the acid and
- the value of the acid dissociation constant, K_a.

Assumptions and approximations

Consider the equilibrium of a weak aqueous acid HA(aq):

$$HA(aq) \rightleftharpoons H^+(aq) + A^-(aq)$$

- Assuming that only a very small proportion of HA dissociates, the equilibrium concentration of HA(aq) will be very nearly the same as the concentration of undissociated HA(aq).

$$\therefore [HA(aq)]_{equilibrium} \approx [HA(aq)]_{start}$$

- Assuming that there is a negligible proportion of $H^+(aq)$ from ionisation of water:

$$[H^+(aq)] \approx [A^-(aq)]$$

- Using these approximations

$$K_a = \frac{[H^+(aq)][A^-(aq)]}{[HA(aq)]} \qquad \therefore K_a \approx \frac{[H^+(aq)]^2}{[HA(aq)]}$$

For calculations, use
$$K_a \approx \frac{[H^+(aq)]^2}{[HA(aq)]}$$

Rates and equilibria

Take care to learn this method.
Many marks are dropped on exam papers by students who have not done so!

Example
For a weak acid $[HA(aq)] = 0.100$ mol dm^{-3}, $K_a = 1.70 \times 10^{-5}$ mol dm^{-3} at 25°C. Calculate the pH.

$$K_a = \frac{[H^+(aq)][A^-(aq)]}{[HA(aq)]} \approx \frac{[H^+(aq)]^2}{[HA(aq)]}$$

$$\therefore 1.70 \times 10^{-5} = \frac{[H^+(aq)]^2}{0.100}$$

$$\therefore [H^+(aq)] = \sqrt{0.100 \times 1.70 \times 10^{-5}} = 0.00130 \text{ mol dm}^{-3}$$

$$pH = -\log_{10}[H^+(aq)] = -\log_{10}(0.00130) = \textbf{2.89}$$

K_a and pK_a

AQA M4

Values of K_a can be made more manageable if expressed in a logarithmic form, pK_a (see also page 27: pH and $[H^+(aq)]$).

$$pK_a = -\log_{10} K_a$$
$$K_a = 10^{-pK_a}$$

K_a and pK_a conversions are just like those between pH and H$^+$.

- A **low** value of K_a matches a **high** value of pK_a
- A **high** value of K_a matches a **low** value of pK_a

The smaller the pK_a value, the stronger the acid.

Comparison of K_a and pK_a

acid		K_a / mol dm^{-3}	pK_a
methanoic acid	HCOOH	1.6×10^{-4}	$-\log_{10}(1.6 \times 10^{-4}) = 3.8$
benzoic acid	C$_6$H$_5$COOH	6.3×10^{-5}	$-\log_{10}(6.3 \times 10^{-5}) = 4.2$

Progress check

1. Find the pH of solutions of a weak acid HA ($K_a = 1.70 \times 10^{-5}$ mol dm^{-3}), with the following concentrations:
   (a) 1.00 mol dm^{-3}; (b) 0.250 mol dm^{-3}; (c) 3.50×10^{-2} mol dm^{-3}

2. Find values of K_a and pK_a for the following weak acids.
   (a) 1.0 mol dm^{-3} solution with a pH of 4.5
   (b) 0.1 mol dm^{-3} solution with a pH of 2.2
   (c) 2.0 mol dm^{-3} solution with a pH of 3.8.

1 (a) 2.38 (b) 2.69 (c) 3.11
2 (a) $K_a = 1 \times 10^{-9}$ mol dm^{-3}, $pK_a = 9$
  (b) $K_a = 3.98 \times 10^{-4}$ mol dm^{-3}, $pK_a = 3.4$
  (c) $K_a = 1.26 \times 10^{-8}$ mol dm^{-3}, $pK_a = 7.9$

The ionisation of water and K_w

AQA M4

In water, a very small proportion of molecules dissociates into H$^+$(aq) and OH$^-$(aq) ions. The position of equilibrium lies well to the left of the equation below, representing this dissociation.

$$H_2O(l) \underset{\text{equilibrium}}{\rightleftharpoons} H^+(aq) + OH^-(aq)$$

Treating water as a weak acid: $K_a = \dfrac{[H^+(aq)][OH^-(aq)]}{[H_2O(l)]}$

Rearranging gives:

$$\underbrace{K_a \times [H_2O(l)]}_{\text{constant, } K_w} = [H^+(aq)][OH^-(aq)]$$

[H$_2$O(l)] is constant and is included within K_w.

> **KEY POINT**
>
> The constant K_w is called the **ionic product of water**
> - $K_w = [H^+(aq)][OH^-(aq)]$
> - At 25°C, $K_w = 1.0 \times 10^{-14}$ mol^2 dm^{-6}.

Rates and equilibria

In water, the concentrations of $H^+(aq)$ and $OH^-(aq)$ ions are the same.

- $[H^+(aq)] = [OH^-(aq)] = 10^{-7}$ mol dm^{-3} ($10^{-14} = 10^{-7} \times 10^{-7}$)

Hydrogen ion and hydroxide ion concentrations

All aqueous solutions contain $H^+(aq)$ and $OH^-(aq)$ ions. The proportions of these ions in a solution are determined by the pH.

In water	$[H^+(aq)] = [OH^-(aq)]$
In **acidic** solutions	$[H^+(aq)] > [OH^-(aq)]$
In **alkaline** solutions	$[H^+(aq)] < [OH^-(aq)]$

The concentrations of $H^+(aq)$ and $OH^-(aq)$ are linked by K_w.

- At 25°C $1.0 \times 10^{-14} = [H^+(aq)][OH^-(aq)]$
- The indices of $[H^+(aq)]$ and $[OH^-(aq)]$ add up to –14.

> **Linking [H$^+$] and [OH$^-$]**
> $10^{-14} = [H^+][OH^-]$
> Water: pH = 7,
> $[H^+] = 10^{-7}$ mol dm^{-3}
> $10^{-14} = 10^{-7} \times 10^{-7}$
> An acid: pH = 3
> $[H^+] = 10^{-3}$ mol dm^{-3}
> $10^{-14} = 10^{-3} \times 10^{-11}$
> An alkali: pH = 10
> $[H^+] = 10^{-10}$ mol dm^{-3}
> $10^{-14} = 10^{-10} \times 10^{-4}$

Calculating the pH of strong alkalis

The pH of a strong alkali can be found using K_w.

For a strong alkali, BOH:

- we can assume complete dissociation
- the concentration of $OH^-(aq)$ can be found directly from the alkali concentration: $[BOH] = [OH^-]$.

Example

A strong alkali, BOH, has a concentration of 0.50 mol dm^{-3}.
What is the pH?

Complete dissociation.   ∴ $[OH^-(aq)] = 0.50$ mol dm^{-3}

$K_w = [H^+(aq)][OH^-(aq)] = 1 \times 10^{-14}$ mol^2 dm^{-6}

∴ $[H^+(aq)] = \dfrac{K_w}{[OH^-(aq)]} = \dfrac{1 \times 10^{-14}}{0.50} = 2 \times 10^{-14}$ mol dm^{-3}

$pH = -\log_{10}[H^+(aq)] = -\log_{10}(2 \times 10^{-14}) = $ **13.7**

> To find the pH of an alkali, first find $[H^+]$ using K_w and $[OH^-]$.

Progress check

1. Find the $[H^+(aq)]$ and pH of the following alkalis at 25°C:
   (a) 1×10^{-3} mol dm^{-3} $OH^-(aq)$
   (b) 3.5×10^{-2} mol dm^{-3} $OH^-(aq)$.

2. Find the pH of the following solutions of strong bases at 25°C:
   (a) 0.01 mol dm^{-3} KOH (aq)
   (b) 0.20 mol dm^{-3} NaOH (aq).

1. (a) $[H^+(aq)] = 1 \times 10^{-11}$ mol dm^{-3}; pH = 11
   (b) $[H^+(aq)] = 2.86 \times 10^{-13}$ mol dm^{-3}; pH = 12.5.
2. (a) pH = 12; (b) pH = 13.3.

Rates and equilibria

1.6 pH changes

After studying this section you should be able to:

- recognise the shapes of titration curves for acids and bases with different strengths
- explain the choice of suitable indicators for acid–base titrations
- determine K_a from a titration curve
- carry out unstructured titration calculations
- state what is meant by a buffer solution
- explain how pH is controlled by each component in a buffer solution
- calculate the pH of a buffer solution

Titration curves

Choosing an indicator using titration curves

A titration curve shows the changes in pH during a titration.

In the titration curves below, different combinations of strong and weak acids and alkalis have been used.

- The end point of a titration is identified by the colour change of the indicator.
- Different indicators change colour at different pH values.
- The pH values at which the indicators methyl orange (**MO**) and phenolphthalein (**P**) change colour are shown on each diagram.

strong acid/strong alkali

both indicators suitable

strong acid/weak alkali

*methyl orange (**MO**) suitable*
*phenolphthalein (**P**) unsuitable*

Choose the correct indicator

Strong acid/strong alkali
phenolphthalein ✓
methyl orange ✓

Strong acid/weak alkali
phenolphthalein ✗
methyl orange ✓

Weak acid/strong alkali
phenolphthalein ✓
methyl orange ✗

Weak acid/weak alkali
phenolphthalein ✗
methyl orange ✗

weak acid/strong alkali

*phenolphthalein (**P**) suitable*
*methyl orange (**MO**) unsuitable*

weak acid/weak alkali

neither indicator suitable

Key features of titration curves

- The pH changes rapidly at the near vertical portion of the titration curve. The mid-point of the vertical portion is the **equivalence point** of the titration.
- The sharp change in pH is brought about by a very small addition of alkali, typically the addition of one drop.
- The indicator is only suitable if its pK_{ind} value is within the pH range of the near vertical portion of the titration curve.

Rates and equilibria

Unstructured titration calculations

AQA M4, M5

In AS chemistry examinations, titration problems are usually highly structured. You are helped through calculations with each step being asked in a sequence that leads to the final solution.

In A2 chemistry examinations:

- problems will be set in which you are only provided with a summary of the experimental results
- you will be expected to navigate your own way through the problem; you will have no hints about how to solve the problem.

The key to success is a good understanding of the stages required to solve a titration problem. Most problems follow a common framework.

Acid–base titrations are always likely to be carried out in chemical analysis. In A2, you will also come across redox titrations (see pages 74–75). Although these are different types of titration, the principles for solving any calculations are the same.

> AS exam questions would help you through titration calculations by asking each step as a question.
> But at A2 you are really on your own! You should now know more chemistry than at AS and also be able to tackle chemistry problems independently.
> At least that is what the setters of examination questions expect!

Stage 1

All titrations use a standard solution – one whose concentration is known. In the titration, you measure the volume of this solution that exactly reacts with a second solution.

- You first find the amount, in moles, of the compound in the standard solution.

E.g. In a titration 25.0 cm³ of 0.100 mol dm⁻³ NaOH reacted exactly with a solution of H_2SO_4.

$$n = c \times \frac{V}{1000} = 0.100 \times \frac{25.0}{1000} = 0.00250 \text{ mol}$$

> Remember that the steps shown here are key to solving most titration calculations.
> See also redox titrations, pages 74–75.

Stage 2

- The equation is then usually used to find the number of moles of the other substance dissolved in the second solution:

E.g. $2NaOH(aq) + H_2SO_4(aq) \longrightarrow$ products

                2 mol   1 mol

amounts: 0.00250   0.00125 mol

Stage 3

- Finally, this value is processed in some way to find out some further information about the second substance. This is the hardest stage and it may contain several steps. This stage may also vary between different problems.

Worked example

Compound **A** is a straight-chain carboxylic acid. A student analysed a sample of acid **A** by the procedure below.

The student dissolved 5.437 g of **A** in water and made the solution up to 250 cm³.

In a titration, 25.0 cm³ of 0.200 mol dm⁻³ NaOH were neutralised by exactly 23.45 cm³ of solution **A**.

Use the results to calculate the molar mass of acid **A** and suggest its identity.

> The concentration and volume of NaOH are known. So we work out the amount of NaOH first.

The amount of NaOH used can be calculated from the titration results:

amount of NaOH = $0.200 \times \frac{25.0}{1000} = 0.00500$ mol

Rates and equilibria

The amount of carboxylic acid that reacts with the NaOH can be determined from the balanced equation for the reaction:

Carboxylic acid **A** must have the formula: RCOOH

The equation and molar reacting quantities are:

$$NaOH(aq) + RCOOH(aq) \longrightarrow RCOONa(aq) + H_2O(l)$$
$$1 \text{ mol} \qquad 1 \text{ mol}$$

∴ in the titration: 0.00500 mol NaOH reacts with **0.00500 mol RCOOH**

> In this titration there is the same number of moles of each reactant. We work this out from the balanced equation.

You must now look at what the question is asking. You need to think about how you are going to solve the rest of the problem.

- Here, we must first find out the amount, in moles, of **A** that was used to prepare the original 250 cm³ solution.
- In the titration, 23.45 cm³ of this solution of **A** was used – we know the number of moles of **A** dissolved in this volume.
- We need to scale up these quantities to find the number of moles of **A** dissolved in 250 cm³ of solution:

23.45 cm³ of RCOOH(aq) in the titration contains 0.00500 mol RCOOH

1 cm³ of RCOOH(aq) in the titration contains $\dfrac{0.00500}{23.45}$ mol RCOOH

250 cm³ of RCOOH(aq) contains $\dfrac{0.00500}{23.45} \times 250 = 0.0533$ mol RCOOH

> This step is important but many candidates forget to do this (or don't understand that it must be done!)

The molar mass of RCOOH can now be determined and we have solved the first part of the problem:

$$n = \frac{m}{M} \quad \therefore \text{ Molar mass, } M = \frac{m}{n} = \frac{5.437}{0.0533} = 102.0 \text{ g mol}^{-1}$$

> This step is easy, provided that you have remembered the following:
> $n = \dfrac{\text{mass } m}{\text{Molar mass } M}$

*Finally, we need to suggest a possible structure for straight-chain carboxylic acid **A**.*

The formula of the straight-chain carboxylic acid is RCOOH.

COOH has a relative mass of 12.0 + 16.0 × 2 + 1.0 = 45.0.

∴ the alkyl group R has a relative mass of 102.0 − 45.0 = 57.0

∴ the alkyl group must be $CH_3CH_2CH_2CH_2$ (15.0 + 14.0 + 14.0 + 14.0)

and carboxylic acid **A** is **$CH_3CH_2CH_2CH_2COOH$** (pentanoic acid).

> These problems may mix together chemistry from different parts of your synoptic assessment. See pages 141–144.

Buffer solutions

AQA — M4

A buffer solution minimises changes in pH during the addition of an acid or an alkali to a solution. The buffer solution maintains a near-constant pH by removing most of any added acid or alkali.

A **buffer solution** is a mixture of:

- a **weak acid**, HA, and
- its **conjugate base**, A^-:

$$HA(aq) \rightleftharpoons H^+(aq) + A^-(aq)$$
$$\text{weak acid} \qquad\qquad\qquad \text{conjugate base}$$

> Buffers are added to foods to prevent deterioration due to pH change (caused by bacterial or fungal activity).

In a buffer solution, the concentration of hydrogen ions, $[H^+(aq)]$, is very small compared with the concentrations of the weak acid $[HA(aq)]$ or the conjugate base $[A^-(aq)]$.

$$[H^+(aq)] \ll [HA(aq)] \quad \text{and} \quad [H^+(aq)] \ll [A^-(aq)].$$

Rates and equilibria

> We can explain how a buffer minimises pH changes by using le Chatelier's principle.

How does a buffer act?

Addition of an acid, H⁺(aq), to a buffer

On addition of an acid:

- $[H^+(aq)]$ is increased
- the pH change is opposed and the **equilibrium moves to the left**, removing $[H^+(aq)]$ and forming HA(aq)
- the **conjugate base** $A^-(aq)$ removes most of any added $[H^+(aq)]$.

> A^- removes most of any added acid.

add acid, H⁺(aq)
↓

| HA(aq) ⇌ H⁺(aq) + A⁻(aq) |
| weak acid   conjugate base |

← most added H⁺(aq) is removed

Addition of an alkali, OH⁻(aq), to a buffer

On addition of an alkali:

- the added $OH^-(aq)$ reacts with the small concentration of $H^+(aq)$:

$$H^+(aq) + OH^-(aq) \longrightarrow H_2O(l)$$

- the pH change is opposed – the **equilibrium moves to the right**, restoring $[H^+(aq)]$ as HA(aq) dissociates
- the **weak acid HA** restores most of any $[H^+(aq)]$ that has been removed.

add alkali, OH⁻(aq)
↑

| HA(aq) ⇌ H⁺(aq) + A⁻(aq) |
| weak acid   conjugate base |

→ most H⁺(aq) is restored

> HA removes added alkali.

> **KEY POINT**
> Although the two components in a buffer solution react with added acid and alkali, they **cannot stop** the pH from changing. They do however **minimise** pH changes.

Common buffer solutions

> Remember these buffers:
> CH₃COOH/CH₃COONa
> and
> NH₄Cl/NH₃

An acidic buffer

A common acidic buffer is an aqueous solution containing a mixture of ethanoic acid, CH_3COOH, and the ethanoate ion, CH_3COO^-.

- Ethanoic acid acts as the **weak acid**, CH_3COOH.
- Sodium ethanoate, $CH_3COO^-Na^+$, acts as a source of the **conjugate base**, CH_3COO^-.

> The pH of healthy blood is maintained at between 7.35 and 7.40 by the carbonic acid–hydrogencarbonate buffer system.
> H₂CO₃ is the **weak acid**,
> HCO₃⁻ is the **conjugate base**.
> H₂CO₃(aq) ⇌ H⁺(aq)+HCO₃⁻(aq)
> Blood cannot sustain life if its pH lies much outside of this narrow range.

An alkaline buffer

A common alkaline buffer is an aqueous solution containing a mixture of the ammonium ion, NH_4^+, and ammonia, NH_3.

- Ammonium chloride, $NH_4^+Cl^-$, acts as source of the **weak acid**, NH_4^+.
- Ammonia acts as the **conjugate base**, NH_3.

Rates and equilibria

Calculations involving buffer solutions

AQA M4

The pH of a buffer solution depends upon:
- the acid dissociation constant, K_a, of the buffer system
- the **ratio** of the weak **acid** and its conjugate **base**.

For a buffer comprising the weak acid, HA, and its conjugate base, A^-,

$$K_a = \frac{[H^+(aq)][A^-(aq)]}{[HA(aq)]}$$

> For buffer calculations, learn:
>
> $[H^+(aq)] = K_a \times \dfrac{[HA(aq)]}{[A^-(aq)]}$
>
> The pH of a buffer can be controlled by the acid/base ratio.

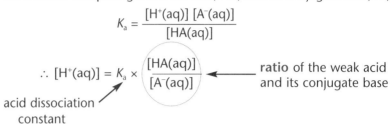

$$\therefore [H^+(aq)] = K_a \times \frac{[HA(aq)]}{[A^-(aq)]}$$

acid dissociation constant — ratio of the weak acid and its conjugate base

Example

Calculate the pH of a buffer comprising 0.30 mol dm^{-3} CH$_3$COOH(aq) ($K_a = 1.7 \times 10^{-5}$ mol dm^{-3}) and 0.10 mol dm^{-3} CH$_3$COO$^-$(aq).

$$[H^+(aq)] = K_a \times \frac{[HA(aq)]}{[A^-(aq)]}$$

$$\therefore [H^+(aq)] = 1.7 \times 10^{-5} \times \frac{0.30}{0.10} = 5.1 \times 10^{-5} \text{ mol dm}^{-3}$$

$$\therefore \text{pH} = -\log_{10}[H^+(aq)] = -\log_{10}(5.1 \times 10^{-5})$$

$$\therefore \text{pH of the buffer solution} = \mathbf{4.3}$$

Progress check

1. Three buffer solutions are made from benzoic acid, C$_6$H$_5$COOH and sodium benzoate, C$_6$H$_5$COONa with the following compositions:

Buffer A: 0.10 mol dm^{-3} C$_6$H$_5$COOH and 0.10 mol dm^{-3} C$_6$H$_5$COONa
Buffer B: 0.75 mol dm^{-3} C$_6$H$_5$COOH and 0.25 mol dm^{-3} C$_6$H$_5$COONa
Buffer C: 0.20 mol dm^{-3} C$_6$H$_5$COOH and 0.80 mol dm^{-3} C$_6$H$_5$COONa

   (a) Write the equation for the equilibrium in these buffers.
   (b) Write an expression for K_a of benzoic acid.
   (c) Calculate the pH of each of the buffer solutions (K_a for C$_6$H$_5$COOH = 6.3 × 10^{-5} mol dm^{-3}.)

1 (a) C$_6$H$_5$COOH ⇌ H$^+$(aq) + C$_6$H$_5$COO$^-$(aq)
(b) $K_a = \dfrac{[H^+(aq)][C_6H_5COO^-(aq)]}{[C_6H_5COOH(aq)]}$
(c) Buffer A: 4.2 Buffer B: 3.7 Buffer C: 4.8

Rates and equilibria

Sample question and model answer

Many cosmetics contain buffers. Human skin is slightly acidic and has a pH of approximately 5.50. Skin-care products often buffer at a pH 5.50 or below.

(a) What do you understand by the term *buffer*?

A solution that minimises a change in pH. ✓ [1]

> Don't use the phrases 'keeps the pH constant' or 'stops the pH from changing'.
> The pH will change but the buffer keeps the change small.

(b) Lactic acid, $CH_3CH(OH)COOH$, is used in many cosmetics. Lactic acid is a weak acid with an acid dissociation constant, K_a, of 8.40×10^{-4} mol dm^{-3}.

(i) Write an equation to show the dissociation of lactic acid into its ions.

$CH_3CH(OH)COOH \rightleftharpoons CH_3CH(OH)COO^- + H^+$ ✓ (equilibrium sign essential)

(ii) Write an expression for the acid dissociation constant, K_a, of lactic acid.

$$K_a = \frac{[CH_3CH(OH)COO^-(aq)]\,[H^+(aq)]}{[CH_3CH(OH)COOH(aq)]}$$ ✓

(iii) Calculate the pH of a 1.25×10^{-2} mol dm^{-3} solution of lactic acid.

$CH_3CH(OH)COO^- \approx H^+$.

equilibrium conc. of weak acid ≈ undissociated concentration of weak acid.

$$\therefore K_a \approx \frac{[H^+(aq)]^2}{[CH_3CH(OH)COOH(aq)]} \qquad 8.4 \times 10^{-4} = \frac{[H^+(aq)]^2}{1.25 \times 10^{-2}}$$ ✓

$[H^+(aq)] = \sqrt{1.25 \times 10^{-2} \times 8.40 \times 10^{-4}} = 3.24 \times 10^{-3}$ ✓ mol dm^{-3}

$\therefore$ pH = $- \log (3.24 \times 10^{-3}) = 2.49$. ✓ [5]

> These approximations simplify the K_a expression to:
> $K_a \approx \frac{[H^+]^2}{[HA]}$

> Make sure that your calculator skills are good and show your working. A correct method will secure most available marks.
> Notice the use throughout of 3 significant figures (in the question and in the answers).

(c) A buffer solution can be made based on lactic acid and a salt of lactic acid. Explain how this buffer solution acts as a buffer.

In the equilibrium $CH_3CH(OH)COOH \rightleftharpoons CH_3CH(OH)COO^- + H^+$, there are large excesses of $CH_3CH(OH)COOH$ and $CH_3CH(OH)COO^-$. ✓

The equilibrium above shifts in response to added acid and alkali. ✓

On addition of an alkali, OH$^-$ ions react with the small concentration of H$^+$ present:

$H^+(aq) + OH^-(aq) \longrightarrow H_2O(l)$

The equilibrium shifts to the right, restoring most of any H$^+$ ions removed: ✓

$CH_3CH(OH)COOH \longrightarrow CH_3CH(OH)COO^- + H^+$ ✓

Lactate ions from sodium lactate remove most of any added acid, shifting the equilibrium to the left: ✓

$CH_3CH(OH)COO^- + H^+ \longrightarrow CH_3CH(OH)COOH$ ✓ [6]

> Practise explaining how a buffer works. This is often asked in exams and it is harder to answer than it appears.
> You need to discuss shifts in the equilibrium between the weak acid and its conjugate base.

(d) A buffer solution based on lactic acid is required to buffer at a pH 3.80. Calculate the ratio of lactic acid : sodium lactate that will be needed to give this pH.

In the buffer, $[H^+(aq)] = 10^{-3.80} = 1.58 \times 10^{-4}$ ✓ mol dm^{-3}

In the buffer, $[H^+(aq)] = K_a \times \frac{[CH_3CH(OH)COOH(aq)]}{[CH_3CH(OH)COO^-(aq)]}$ ✓

$\therefore$ ratio $\frac{[CH_3CH(OH)COOH(aq)]}{[CH_3CH(OH)COO^-(aq)]} = \frac{[H^+(aq)]}{K_a} = \frac{1.58 \times 10^{-4}}{8.4 \times 10^{-4}}$ ✓

ratio lactic acid : sodium lactate = $0.189 = \frac{1}{5.30}$ ✓ [4]

[Total: 16]

> The calculated ratio is 0.189.
> To get $\frac{1}{5.30}$ you need to press the x^{-1} key.
> The answer means that you need 5.3 times the concentration of sodium lactate to lactic acid.

Rates and equilibria

Practice examination questions

1 The reaction between hydrogen peroxide and iodide ions in an acid solution can be written as follows.

$$H_2O_2(aq) + 2I^-(aq) + 2H^+(aq) \longrightarrow I_2(aq) + 2H_2O(l)$$

(a) The table below shows initial rates from different initial concentrations of $H_2O_2(aq)$, $I^-(aq)$ and $H^+(aq)$ at constant temperature.

$[H_2O_2(aq)]$ /mol dm^{-3}	$[I^-(aq)]$ /mol dm^{-3}	$[H^+(aq)]$ /mol dm^{-3}	initial rate /10^{-6} mol dm^{-3} s^{-1}
0.00075	0.10	0.10	2.1
0.00150	0.10	0.10	4.2
0.00150	0.10	0.20	4.2
0.00075	0.70	0.10	14.7

   (i) Determine, with reasoning, the order with respect to each reactant.

   (ii) Write the rate equation and calculate the value of the rate constant, k, for this reaction. [9]

(b) (i) What is meant by the term *rate-determining step*?

   (ii) Using the rate equation in (a)(ii), suggest an equation for the rate-determining step of this reaction. [2]

[Total: 11]

2 (a) A chemical reaction is second order with respect to compound **A** and first order with respect to compound **B**.

   (i) Write the rate equation for this reaction.

   (ii) What is the overall order of this reaction?

   (iii) By what factor will the rate increase if the concentrations of **A** and **B** are both tripled?

   (iv) When [**A**] = 0.40 mol dm^{-3} and **B** = 0.35 mol dm^{-3}, the rate = 1.76 x 10^{-4} mol dm^{-3} s^{-1}. Calculate the value of the rate constant, stating its units. [5]

(b) Propanone and iodine react together in acidic solution to give iodopropanone according to the overall equation below.

$$CH_3COCH_3(aq) + I_2(aq) \longrightarrow CH_3COCH_2I(aq) + HI(aq)$$

The rate equation is: $rate = k[CH_3COCH_3(aq)] [H^+(aq)]$

   (i) State, with an explanation, the role of the $H^+(aq)$ ions in this reaction.

   (ii) Suggest a possible equation for the rate-determining step. [3]

[Total: 8]

3 The equilibrium below was set up.

$$CH_3COOCH_2CH_3 + H_2O \rightleftharpoons CH_3COOH + CH_3CH_2OH$$

The equilibrium mixture contains 1.35 mol of ethyl ethanoate, 1.35 mol of water, 0.15 mol of ethanoic acid, and 3.15 mol of ethanol. It has a volume of 353 cm^3.

(a) Calculate K_c for this equilibrium to 2 significant figures. State the units, if any. [3]

(b) More ethanol was to be added to the equilibrium mixture.

   (i) What would happen to the equilibrium? Explain your reasoning.

   (ii) What would be the effect of this change on the value of K_c? Explain your reasoning. [4]

[Total: 7]

37

Rates and equilibria

Practice examination questions (continued)

4 When 0.60 moles of $H_2(g)$ and 0.18 moles of $I_2(g)$ were heated to constant temperature in a sealed container with a volume of 1 dm³, an equilibrium was set up:

$$H_2(g) + I_2(g) \rightleftharpoons 2HI(g)$$

At equilibrium, 0.16 mol $I_2(g)$ had reacted.

(a) (i) Determine the equilibrium concentrations of H_2, I_2, and HI.

(ii) Calculate the equilibrium constant, K_c, for this reaction. [6]

(b) When the equilibrium mixture was heated by 100°C the value of K_c decreased. What additional information is provided by this observation? [2]

[Total: 8]

5 Nitric acid, HNO_3 is a strong acid and methanoic acid, HCOOH, is a weak acid ($K_a = 1.6 \times 10^{-4}$ mol dm⁻³).

(a) Define the term *Brønsted–Lowry acid*. [1]

(b) Write the acid–base equilibrium that would be set up between methanoic acid and nitric acid. State the role of methanoic acid in this equilibrium. [2]

(c) Calculate the pH of the following solutions at 25°C:

(i) 0.176 mol dm⁻³ nitric acid

(ii) 0.372 mol dm⁻³ potassium hydroxide

(iii) 0.263 mol dm⁻³ methanoic acid.

(iv) a solution containing 0.255 mol dm⁻³ methanoic acid and 0.325 sodium methanoate. [10]

[Total: 13]

6 (a) (i) What is meant by the terms *strong acid* and *weak acid*?

(ii) Define the term *Brønsted–Lowry base*. [2]

(b) 25.0 cm³ of a solution of sodium hydroxide contained 1.50×10^{-3} mol of NaOH.

(i) Calculate the pH of this solution at 298 K.

(ii) A student added 50.0 cm³ of 0.250 mol dm⁻³ HCl to the 25.0 cm³ solution of sodium hydroxide.

Calculate the pH of the solution formed. [8]

(c) A sample of hydrochloric acid had a pH of 1.52. A 25 cm³ sample of this acid was diluted with 15 cm³ of water.

Calculate the pH of the resulting solution. [4]

[Total: 14]

Chapter 2
Energy changes in chemistry

The following topics are covered in this chapter:

- Enthalpy changes
- Entropy changes
- Electrochemical cells
- Predicting redox reactions

2.1 Enthalpy changes

After studying this section you should be able to:

- explain and use the term 'lattice enthalpy'
- construct Born–Haber cycles to calculate the lattice enthalpy of simple ionic compounds
- explain the effect of ionic charge and ionic radius on the numerical magnitude of a lattice enthalpy
- calculate enthalpy changes of solution for ionic compounds from enthalpy changes of hydration and lattice enthalpies
- calculate enthalpy changes from mean bond enthalpies.

Key points from AS

- Enthalpy changes
  Revise AS pages 74–82

During the study of enthalpy changes in AS Chemistry, you learnt how to:

- calculate enthalpy changes directly from experiments using the relationship $Q = mc\Delta T$
- calculate enthalpy changes indirectly using Hess' law.

These key principles are built upon by considering the enthalpy changes that bond together an ionic lattice.

Lattice enthalpy

AQA M5

Ionic bonding is the electrostatic attraction between oppositely charged ions. Lattice enthalpy indicates the strength of the ionic bonds in an ionic lattice.

Key points from AS

- Ionic bonding
  Revise AS pages 40–44

> **KEY POINT**
> The lattice enthalpy ($\Delta H^\ominus_{L.E.}$) of an ionic compound is the enthalpy change that accompanies the formation of 1 mole of an ionic compound from its constituent gaseous ions. ($\Delta H^\ominus_{L.E.}$ is **exothermic**.)

$$Na^+(g) + Cl^-(g) \longrightarrow Na^+Cl^-(s)$$

Each ion is surrounded by oppositely-charged ions, forming a giant ionic lattice.

Lattice enthalpy is also sometimes called the **enthalpy change of lattice formation**. The opposite change, to break up the lattice, is called the **enthalpy change of lattice dissociation**.

$$Na^+Cl^-(s) \longrightarrow Na^+(g) + Cl^-(g)$$

Determination of lattice enthalpies

AQA M5

Lattice enthalpies cannot be determined directly and must be calculated indirectly using Hess' Law from other enthalpy changes that can be found experimentally. The energy cycle used to calculate a lattice enthalpy is the **Born–Haber cycle**.

The basis of the Born–Haber cycle is the formation of an ionic lattice from its element by two routes.

Energy changes in chemistry

The Born–Haber cycle for sodium chloride is shown below.

Definitions for these enthalpy changes are shown below.

Route 1: A + B + C + D + E   Route 2: F

Using Hess' Law: A + B + C + D + E = F

$\Delta H^\ominus_{at} Na(g) + \Delta H^\ominus_{at} Cl(g) + \Delta H^\ominus_{I.E.} Na(g) + \Delta H^\ominus_{E.A.} Cl(g) + E = \Delta H^\ominus_f Na^+Cl^-(s)$

∴ +107 + 122 + 496 + (−349) + E = −411

Hence, the lattice energy of NaCl(s), **E = −787 kJ mol⁻¹**

- the 1st ionisation energy of Na is **endothermic**
- the 1st electron affinity of Cl is **exothermic**.

Definitions for enthalpy changes

The enthalpy changes involved in these two routes are shown below.

> **The standard enthalpy change of formation ($\Delta H^\ominus_f$)** is the enthalpy change that takes place when one mole of a compound in its standard state is formed from its constituent elements in their standard states under standard conditions.
>
> Na(s) + ½Cl₂(g) ⟶ NaCl(s) $\Delta H^\ominus_f = -411$ kJ mol⁻¹

KEY POINT

Be careful when using $\Delta H^\ominus_{at}$ involving diatomic elements.

$\Delta H^\ominus_{at}$ relates to the formation of 1 mole of atoms. For chlorine, this involves ½Cl₂ only.

For gaseous molecules $\Delta H^\ominus_{at}$ can be determined from the **bond enthalpy (B.E.)**.

Cl–Cl(g) ⟶ 2Cl(g)
$\Delta H^\ominus_{B.E.} = +244$ kJ mol⁻¹
½Cl–Cl(g) ⟶ Cl(g)
$\Delta H^\ominus_{at} = +122$ kJ mol⁻¹

> **The standard enthalpy change of atomisation ($\Delta H^\ominus_{at}$)** of an element is the enthalpy change that accompanies the formation of 1 mole of gaseous atoms from the element in its standard state.
>
> Na(s) ⟶ Na(g) $\Delta H^\ominus_{at} = +107$ kJ mol⁻¹
> ½Cl₂(g) ⟶ Cl(g) $\Delta H^\ominus_{at} = +122$ kJ mol⁻¹

KEY POINT

> **The first ionisation energy ($\Delta H^\ominus_{I.E.}$)** of an element is the enthalpy change that accompanies the removal of 1 electron from each atom in 1 mole of gaseous atoms to form 1 mole of gaseous 1+ ions.
>
> Na(g) ⟶ Na⁺(g) + e⁻ $\Delta H^\ominus_{I.E.} = +496$ kJ mol⁻¹

KEY POINT

Be careful when using electron affinities. $\Delta H^\ominus_{E.A.}$ relates to the formation of 1 mole of 1− ions. For chlorine, this involves Cl(g) only.

> **The first electron affinity ($\Delta H^\ominus_{E.A.}$)** of an element is the enthalpy change that accompanies the addition of 1 electron to each atom in 1 mole of gaseous atoms to form 1 mole of gaseous 1− ions.
>
> Cl(g) + e⁻ ⟶ Cl⁻(g) $\Delta H^\ominus_{E.A.} = -349$ kJ mol⁻¹

KEY POINT

Energy changes in chemistry

Factors affecting the size of lattice enthalpies

The strength of an ionic lattice and the value of its lattice enthalpy depend upon:
- ionic size
- ionic charge.

Effect of ionic size

The effect of increasing ionic size can be seen by comparing the lattice enthalpies of sodium halides as shown below.

> Lattice energy has a negative value. You should use the term *'becomes less/more negative'* instead of *'becomes bigger/smaller'* to describe any trend in lattice energy.

compound	lattice enthalpy / kJ mol^{-1}	ions	effect of size of halide ion
NaCl	−787	(+)(−)	ionic size increases:
NaBr	−751	(+)(−)	• charge density decreases • attraction between ions decreases
NaI	−705	(+)(−)	• lattice energy becomes less negative.

Effect of ionic charge

The strongest ionic lattices with the most negative lattice enthalpies contain **small, highly charged ions**.

The diagram below compares the change in ionic size and ionic charge across Period 3 in the Periodic Table.

| Na$^+$ | Mg^{2+} | Al^{3+} | P^{3-} | S^{2-} | Cl$^-$ |

With increased charge on cation, two factors increase the magnitude of lattice energy:
- increased charge produces **more** attraction
- decreasing size produces **more** attraction.

With increased charge on anion, there are two competing factors:
- from Cl$^-$ to P^{3-}, the increasing ionic charge produces **more** attraction
- however, the ionic size also increases producing **less** attraction.

Limitations of lattice enthalpies

Theoretical lattice enthalpies can be calculated by considering each ion as a perfect sphere. Any difference between this theoretical lattice enthalpy and lattice enthalpy calculated using a Born–Haber cycle indicates a degree of covalent bonding caused by polarisation. Polarisation is greatest between:
- a small densely charged cation and
- a large anion.

From the ions shown in the diagram above, we would expect the largest degree of covalency between Al^{3+} and P^{3-}.

Enthalpy change of solution

An ionic lattice dissolves in **polar** solvents (e.g. water). In this process, the giant ionic lattice is broken up by polar water molecules which surround each ion in solution.

When sodium chloride is dissolved in water, the enthalpy change of this process can be measured directly as the enthalpy change of solution.

> **Key points from AS**
> - Indirect determination of enthalpy changes
>   *Revise AS pages 78–79*

Energy changes in chemistry

The energy cycle

The complete energy cycle for dissolving sodium chloride in water is shown below. Definitions of these enthalpy changes are shown below.

Lattice enthalpy **forms** the ionic lattice. Notice that the energy change to break 1 mole of the ionic lattice = – (lattice enthalpy).

The production of hydrated ions comprises two changes:
- hydration of $Na^+(g)$
- hydration of $Cl^-(g)$.

Route 1:   A + B + C
Route 2:   D

Using Hess' Law: A + B + C = D

To determine the enthalpy change of hydration of $Cl^-(aq)$, **C**,

$$-\Delta H^\ominus_{L.E.} NaCl(s) + \Delta H^\ominus_{hyd} Na^+(g) + C = \Delta H^\ominus_{solution} NaCl(s)$$

$$\therefore -(-787) + (-390) + C = +13$$

Hence, the enthalpy change of hydration of $Cl^-(aq)$, **C** = -384 kJ mol^{-1}

Definitions for enthalpy changes

The enthalpy changes involved in these two routes are shown below.

> **KEY POINT**
> The **standard enthalpy change of solution** ($\Delta H^\ominus_{solution}$) is the enthalpy change that accompanies the dissolving of 1 mole of a solute in a solvent to form an infinitely dilute solution under standard conditions.
> $NaCl(s) + aq \longrightarrow Na^+(aq) + Cl^-(aq)$ $\Delta H^\ominus_{solution} = +13$ kJmol^{-1}

> **KEY POINT**
> The **standard enthalpy change of hydration** ($\Delta H^\ominus_{hyd}$) of an ion is the enthalpy change that accompanies the hydration of 1 mole of gaseous ions to form 1 mole of hydrated ions in an infinitely dilute solution under standard conditions.
> $Na^+(g) + aq \longrightarrow Na^+(aq)$ $\Delta H^\ominus_{hyd} = -390$ kJmol^{-1}
> $Cl^-(g) + aq \longrightarrow Cl^-(aq)$ $\Delta H^\ominus_{hyd} = -384$ kJmol^{-1}

Factors affecting the magnitude of hydration enthalpy

The values obtained for each ion depend upon the size of the charge and the size of the ion.

ion	Na^+	Mg^{2+}	Al^{3+}	Cl^-	Br^-	I^-
$\Delta H^\ominus_{hyd}$/kJ mol^{-1}	–390	–1891	–4613	–384	–351	–307
effect of charge	• increasing ionic charge • greater attraction for water					
ionic radius/nm	0.102	0.072	0.053	0.180	0.195	0.215
effect of size	• decreasing ionic size • greater attraction for water			• increasing ionic size • less attraction for water		

Energy changes in chemistry

Bond enthalpy

AQA M5

Enthalpy is stored within chemical bonds and **bond enthalpy** indicates the strength of a chemical bond in a gaseous molecule. For simple molecules, such as $H_2(g)$ and $HCl(g)$, bond enthalpy applies to the following processes:

$H-H(g) \longrightarrow 2H(g)$ $\Delta H = +436$ kJ mol^{-1}
$H-Cl(g) \longrightarrow H(g) + Cl(g)$ $\Delta H = +432$ kJ mol^{-1}

Bond enthalpies are **positive** and refer to **bond breaking** – this process requires energy.

> **Bond enthalpy** is the enthalpy change required to **break** and separate **1 mole of bonds** in the molecules of a gaseous element or compound so that the resulting gaseous species exert no forces upon each other.
>
> **KEY POINT**

Average bond enthalpies

Bond enthalpies such as those above (H–H and H–Cl) apply to specific compounds. Only H_2 can have H–H bonds and the H–H bond enthalpy shown above has a definite value. However, some bonds can have different strengths in different environments.

Not all C–H bonds are created equal.

The Cl atom in CH_3Cl affects the environment of the C–H bonds

C–H bonds have different strengths and different bond enthalpies

Data books provide an indication of the likely bond enthalpy of a particular bond by listing **average** or **mean bond enthalpies**.

An *average* bond enthalpy indicates the strength of a *typical* bond.
The average bond enthalpy for the C–H bond is +413 kJ mol^{-1}.

> A **mean bond enthalpy** gives the typical value of a particular bond, having taken into account the same bond in different environments.
> In particular environments a bond may have a different value for the bond enthalpy than the mean value.
>
> **KEY POINT**

Bond making and bond breaking

Bond breaking requires energy: ENDOTHERMIC.
Bond making releases energy: EXOTHERMIC.

Chemical reactions involve bond breaking followed by bond making.

- Energy is first needed to break bonds in the reactants.
  Bond breaking is an endothermic process and **requires** energy.
- Energy is then released as new bonds are formed in the products.
  Bond making is an exothermic process and **releases** energy.

Bond enthalpy is an endothermic change (ΔH +ve) for bonds being broken.

When bonds are made, the enthalpy change will be the same magnitude but the opposite sign (ΔH –ve).

Using bond enthalpies to determine enthalpy changes

The enthalpy change for a reaction involving simple gaseous molecules can be determined using average bond enthalpies in an **energy cycle**:

- Enthalpy required to break bonds = Σ(bond enthalpies in reactants)
- Enthalpy released to make bonds = $-\Sigma$(bond enthalpies in products)

Notice that the relative strengths of the bonds in the reactants and the bonds in the products decide whether a reaction is exothermic or endothermic.

> $\Delta H = \Sigma$(bond enthalpies in reactants) $- \Sigma$(bond enthalpies in products).
>
> **KEY POINT**

Energy changes in chemistry

For the reaction: $CH_4(g) + 2O_2(g) \longrightarrow CO_2(g) + 2H_2O(g)$,

H—C(H)(H)—H + O=O + O=O ⟶ O=C=O + H-O-H + H-O-H

	4 (C–H) + 2 (O=O)	2 (C=O) + 4 (O–H)
$\Delta H/\text{kJ mol}^{-1}$	(4 × 413) + (2 × 497)	(2 × 805) + (4 × 463)
	Bonds broken: (endothermic)	Bonds made: (exothermic)

$\Delta H = \Sigma$(bond enthalpies in reactants) – Σ(bond enthalpies in products)

∴ ΔH = [(4 × 413) + (2 × 497)] – [(2 × 805) + (4 × 463)] kJ mol^{-1}
= **–816 kJ mol^{-1}**

Note that the calculated value is only approximate – the actual bond enthalpies involved may differ from the average values.

- Bonds with small bond enthalpies will break first.
- Low bond enthalpies indicate that a reaction will take place quickly.

Progress check

1 Using the information below:
   (a) name each enthalpy change **A** to **E**
   (b) construct a Born–Haber cycle for sodium bromide and calculate the lattice enthalpy (enthalpy change of lattice formation) of sodium bromide.

enthalpy change	equation	$\Delta H^\ominus/\text{kJ mol}^{-1}$
A	Na(s) + ½Br$_2$(l) ⟶ NaBr(s)	–361
B	Br(g) + e$^-$ ⟶ Br$^-$(g)	–325
C	Na(s) ⟶ Na(g)	+107
D	Na(g) ⟶ Na$^+$(g) + e$^-$	+496
E	½Br$_2$(l) ⟶ Br(g)	+112

2 You are provided with the following enthalpy changes.

Na$^+$(g) + F$^-$(g) ⟶ NaF(s) $\Delta H = -918$ kJ mol^{-1}
Na$^+$(g) + aq ⟶ Na$^+$(aq) $\Delta H = -390$ kJ mol^{-1}
F$^-$(g) + aq ⟶ F$^-$(aq) $\Delta H = -457$ kJ mol^{-1}

(a) Name each of these three enthalpy changes.
(b) Write an equation, including state symbols, for which the enthalpy change is the enthalpy change of solution of NaF.
(c) Calculate the enthalpy change of solution of NaF.

1 (a) A enthalpy change of formation of sodium bromide
      B 1st electron affinity of bromine
      C enthalpy change of atomisation of sodium
      D 1st ionisation energy of sodium
      E enthalpy change of atomisation of bromine
   (b) –751 kJ mol^{-1}
2 (a) lattice enthalpy of NaF
      enthalpy change of hydration of Na$^+$
      enthalpy change of hydration of F$^-$
   (b) NaF(s) ⟶ Na$^+$(aq) + F$^-$(aq)
   (c) +71 kJ mol^{-1}

Energy changes in chemistry

2.2 Entropy changes

After studying this section you should be able to:

- explain that entropy is a measure of disorder
- calculate entropy changes from standard entropies
- understand that entropy, enthalpy and temperature determine the feasibility of a reaction
- apply rates and equilibrium to industrial applications

Entropy

At 0 K, perfect crystals have zero entropy.

What is entropy?

Entropy, S, is a measure of disorder.

Entropy increases whenever particles become more disordered, e.g.

- when a gas spreads spontaneously through a room
- with increasing temperature
- during changes in state: solid ⟶ liquid ⟶ gas
- when a solid lattice dissolves
  $$NaCl(aq) + aq \longrightarrow Na^+(aq) + Cl^-(aq)$$
- in a reaction in which there is an increase in the number of gaseous molecules (i.e. when a gas is evolved)
  $$NaHCO_3(s) + HCl(aq) \longrightarrow NaCl(aq) + H_2O(l) + CO_2(g)$$
  $$MgCO_3(s) \longrightarrow MgO(s) + CO_2(g)$$

Calculating entropy changes

The standard entropy, $S^\ominus$, of a substance is the entropy content of one mole of a substance, under standard conditions.

The entropy change of a reaction can be calculated using standard entropies of the reactants and products.

> $\Delta S = \Sigma S^\ominus$ (products) $- \Sigma S^\ominus$ (reactants)

Notice that standard entropies have units of J K⁻¹ mol⁻¹.

Example

The equation below is for the carbon reduction of chromium oxide, Cr_2O_3.

$$Cr_2O_3(s) + 3C(s) \longrightarrow 2Cr(s) + 3CO(g)$$

	$Cr_2O_3(s)$	$C(s)$	$Cr(s)$	$CO(g)$
$S^\ominus / J\ K^{-1}\ mol^{-1}$	+81	+6	+24	+198

$\Delta S = \Sigma S^\ominus$ (products) $- \Sigma S^\ominus$ (reactants)
$= [(2 \times 24) + (3 \times 198)] - [(+81) + (3 \times 6)] = 543\ J\ K^{-1}\ mol^{-1}$

Energy changes in chemistry

Spontaneous processes

A **spontaneous** process is one that proceeds of its **own accord**.

Processes lead to **increased stability** if they lead to a **lower energy** state.

This is obviously the situation in an exothermic reaction in which the enthalpy (heat energy) decreases during a reaction. It therefore seems surprising that **some** reactions are **endothermic** and that these can **occur spontaneously** at room temperature.

Enthalpy changes alone cannot explain whether a reaction takes place.

For this, we must consider **both** the **ΔH and ΔS**.

Free energy

AQA M5

Whether a chemical or physical process takes place is dependent on
- the temperature, T,
- the entropy change in the system, ΔS,
- the enthalpy change, ΔH, with the surroundings.

The change in *Free Energy*, ΔG, automatically determines the balance between enthalpy change, entropy change and temperature:

$$\Delta G = \Delta H - T\Delta S$$

> The balance between entropy and enthalpy is the feasibility of a reaction.

Feasibility of a reaction

Processes are spontaneous (feasible) when ΔG is negative. The sign of ΔG depends upon ΔH and ΔS, as shown below.

ΔH	ΔS	ΔG
–ve	+ve	always –ve: reaction feasible
+ve	–ve	always +ve: reaction never feasible
–ve	–ve	–ve at low temperatures
+ve	+ve	–ve at high temperatures

- $\Delta G = \Delta H - T\Delta S$
- The reaction is feasible when $\Delta G < 0$.

> ΔG varies with temperature: when ΔH and ΔS have the same sign, the feasibility of the process depends on the temperature, T.

> Be careful with units:
> ΔH: kJ mol^{-1}
> ΔS: J K^{-1} mol^{-1}

Example

The equation below is for the carbon reduction of chromium oxide, Cr_2O_3.

$$Cr_2O_3(s) + 3C(s) \longrightarrow 2Cr(s) + 3CO(g)$$

$\Delta H = +807$ kJ mol^{-1}; $\Delta S = 543$ J K^{-1} mol^{-1}

What is the minimum temperature that this reaction takes place spontaneously?

$\Delta S = 543$ J K^{-1} mol^{-1} = 0.543 kJ K^{-1} mol^{-1}

$\Delta G = \Delta H - T\Delta S$;

Minimum temperature when $\Delta G = 0$.

$\therefore 0 = \Delta H - T\Delta S$; $T = \dfrac{\Delta H}{\Delta S} = \dfrac{807}{0.543} = 1486$ K

Energy changes in chemistry

Progress check

1. State whether the equations below have a positive or negative entropy change.
   (a) $H_2O(g) \longrightarrow H_2O(l)$
   (b) $N_2(g) + 3H_2(g) \longrightarrow 2NH_3(g)$
   (c) $CaCl_2(s) + aq \longrightarrow Ca^{2+}(aq) + 2Cl^-(aq)$

2. The equation below is for the thermal decomposition of calcium carbonate.
   $$CaCO_3(s) \longrightarrow CaO(s) + CO_2(g) \qquad \Delta H = +178 \text{ kJ mol}^{-1}$$

	$CaCO_3(s)$	$CaO(s)$	$CO_2(g)$
$S^\ominus$ / J K^{-1} mol^{-1}	+93	+40	+214

   (a) What is ΔS?
   (b) What is the minimum temperature for decomposition to take place?

1 (a) –ve; (b) –ve; (c) +ve.
2 (a) 161 J K^{-1} mol^{-1}; (b) 1106 K

Energy changes in chemistry

2.3 Electrochemical cells

After studying this section you should be able to:

- understand the principles of an electrochemical cell
- define the term 'standard electrode (redox) potential', $E^\ominus$
- describe how to measure standard electrode potentials
- calculate a standard cell potential from standard electrode potentials
- write cell diagrams
- describe applications of electrochemical cells, including fuel cells

Key points from AS

- Redox reactions
  Revise AS pages 33–35

During the study of redox in AS Chemistry you learnt:

- that oxidation and reduction involve transfer of electrons in a redox reaction
- the rules for assigning oxidation states
- how to construct an overall redox equation from two half-equations.

These key principles are built upon by considering electrochemical cells in which redox reactions provide electrical energy.

Electricity from chemical reactions

AQA M5

In a redox reaction, electrons are transferred between the reacting chemicals.

For example, zinc reacts with aqueous copper(II) ions in a redox reaction.

$$Zn(s) + Cu^{2+}(aq) \longrightarrow Zn^{2+}(aq) + Cu(s)$$
oxidation: $Zn(s) \longrightarrow Zn^{2+}(aq) + 2e^-$
reduction: $Cu^{2+}(aq) + 2e^- \longrightarrow Cu(s)$

Electrochemical cells are the basis of all batteries. They are used as a portable form of electricity.

If this reaction is carried out simply by mixing zinc with aqueous copper(II) ions:

- heat energy is produced and, unless used as useful heat, is often wasted.

An electrochemical cell controls the transfer of electrons:

- electrical energy is produced and can be used for electricity.

The zinc-copper electrochemical cell

An electrochemical cell comprises of two **half-cells**. In each half-cell, the **oxidation** and **reduction** processes take place **separately**.

A half-cell contains the oxidised and reduced species from a half-equation.

The species can be converted into one another by addition or removal of electrons.

A zinc-copper cell has two half-cells:

- an oxidation half-cell containing reactants and products from the oxidation half-reaction: $Zn^{2+}(aq)$ and $Zn(s)$
- a reduction half-cell containing reactants and products from the reduction half-reaction: $Cu^{2+}(aq)$ and $Cu(s)$.

The circuit is completed using:

- a connecting **wire** which transfers **electrons** between the half-cells
- a **salt bridge** which transfers **ions** between the half-cells.

The salt bridge consists of filter paper soaked in aqueous KNO_3 or NH_4NO_3.

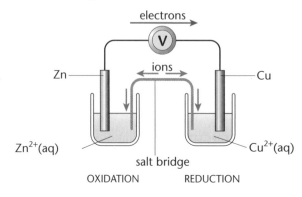

Remember these charge carriers:

***ELECTRONS** through the wire;*

***IONS** through the salt bridge.*

48

oxidation half-cell: Zn^{2+}(aq) and Zn(s)
- **Zn is oxidised** to Zn^{2+}: $\quad Zn(s) \longrightarrow Zn^{2+}(aq) + 2e^-$
- electrons are **supplied** to the external circuit
- the polarity is **negative**.

reduction half-cell: Cu^{2+}(aq) and Cu(s)
- **Cu^{2+} is reduced** to Cu: $\quad Cu^{2+}(aq) + 2e^- \longrightarrow Cu(s)$
- electrons are **taken** from the external circuit
- the polarity is **positive**.

The measured cell e.m.f. of 1.10 V indicates that there is a potential difference of 1.10 V between the copper and zinc half-cells.

Standard electrode potentials

A **hydrogen half-cell** is used as the standard for the measurement of standard electrode potentials.

The standard hydrogen half-cell is based upon the half-reaction below.

$$H^+(aq) + e^- \rightleftharpoons \tfrac{1}{2}H_2(g)$$

A hydrogen half-cell typically comprises

- 1 mol dm^{-3} hydrochloric acid as the source of H^+(aq)
- a supply of hydrogen gas, H_2(g) at 100 kPa
- an inert platinum electrode on which the half-reaction above takes place.

> **KEY POINT**
> The **standard electrode potential** of a half-cell, $E^\ominus$, is the e.m.f. of a half-cell compared with a standard hydrogen half-cell.
> All measurements are at **298 K** with solution **concentrations of 1 mol dm^{-3}** and gas **pressures of 100 kPa**.

Measuring standard electrode potentials

Half-cells comprising a metal and a metal ion

The diagram below shows how a hydrogen half-cell can be used to measure the standard electrode potential of a copper half-cell.

A high-resistance voltmeter is used to minimise the current that flows.

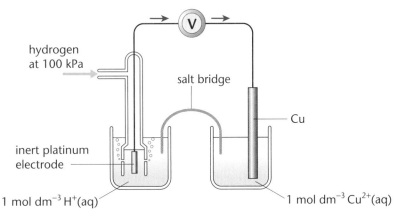

Hydrogen half-cell

> **KEY POINT**
> - The standard electrode potential of a half-cell represents its contribution to the cell e.m.f.
> - The contribution made by the hydrogen half-cell to the cell e.m.f. is defined as 0 V.
> - The **sign** of the standard electrode potential of a half-cell indicates its **polarity** compared with the hydrogen half-cell.

Energy changes in chemistry

As with the hydrogen half-cell, the platinum electrode provides a surface on which the half-reaction can take place.

Half-cells comprising ions of different oxidation states

The half-reaction in a half-cell can be between aqueous ions of the same element with different oxidation states. To allow electrons to pass into the half-cell, an inert electrode of platinum is used.

A half-cell can contain $Fe^{3+}(aq)$ and $Fe^{2+}(aq)$ ions:

$$Fe^{3+}(aq) \rightleftharpoons Fe^{2+}(aq) + e^-$$

A **standard** $Fe^{3+}(aq)/Fe^{2+}(aq)$ half-cell requires:

Alternatively the solution can contain **equal** concentrations of Fe^{2+} and Fe^{3+}.

- a solution containing **both**
  - 1 mol dm^{-3} $Fe^{3+}(aq)$ **and**
  - 1 mol dm^{-3} $Fe^{2+}(aq)$
- an inert platinum electrode with a connecting wire.

The electrochemical series

A standard electrode potential indicates the availability of electrons from a half-reaction.

Metals are reducing agents.

- Metals typically react by donating electrons in a redox reaction.
- Reactive metals have the greatest tendency to donate electrons and their half-cells have the most negative electrode potentials.

Non-metals are oxidising agents.

Key points from AS

- **Reactivity of s-block elements**
  Revise AS page 61
- **The relative reactivity of the halogens as oxidising agents**
  Revise AS page 63

- Non-metals typically react by accepting electrons in a redox reaction.
- Reactive non-metals have the greatest tendency to accept electrons and their half-cells have the most positive electrode potentials.

The **electrochemical series** shows electrode reactions listed in order of their $E^\ominus$ values. The electrochemical series below lists with the **most positive** $E^\ominus$ value at the top. This enables oxidising and reducing agents to be easily compared. The reduced form is shown on the right-hand side of the electrode equilibrium.

This version of the electrochemical series shows standard electrode potentials listed with the **most positive** $E^\ominus$ value at the top.

You will find versions of the electrochemical series listed with the **most negative** $E^\ominus$ value at the top.

This does not matter. The important point is that the potentials are listed **in order** and **can be compared**.

	electrode reaction				$E^\ominus$/ V	
strongest oxidising agent	$F_2(g)$ +	$2e^-$	$\rightleftharpoons$	$2F^-(aq)$	+2.87	
	$Cl_2(g)$ +	$2e^-$	$\rightleftharpoons$	$2Cl^-(aq)$	+1.36	
	$Br_2(l)$ +	$2e^-$	$\rightleftharpoons$	$2Br^-(aq)$	+1.09	
	$Ag^+(aq)$ +	e^-	$\rightleftharpoons$	$Ag(s)$	+0.80	
	$Cu^{2+}(aq)$ +	$2e^-$	$\rightleftharpoons$	$Cu(s)$	+0.34	
	$H^+(aq)$ +	e^-	$\rightleftharpoons$	$\frac{1}{2}H_2(g)$	0	
	$Fe^{2+}(aq)$ +	$2e^-$	$\rightleftharpoons$	$Fe(s)$	−0.44	
	$Zn^{2+}(aq)$ +	$2e^-$	$\rightleftharpoons$	$Zn(s)$	−0.76	strongest reducing agent
	$Cr^{3+}(aq)$ +	$3e^-$	$\rightleftharpoons$	$Cr(s)$	−0.77	
	$K^+(aq)$ +	e^-	$\rightleftharpoons$	$K(s)$	−2.92	

> **KEY POINT**
> In the electrochemical series:
> - the **oxidised** form is on the **left-hand side** of the half-equation
> - the **reduced** form is on the **right-hand side** of the half-equation.

Standard cell potentials and cell reactions

The **standard cell potential** of a cell is the e.m.f. acting between the two half-cells making up the cell under standard conditions.

The **cell reaction** is:

Key points from AS

- **Combining half-equations**
  Revise AS page 35

- the overall process taking place in the cell
- the sum of the reduction and oxidation half-reactions taking place in each half-cell.

50

Energy changes in chemistry

Calculating the standard cell potential of a silver-iron(II) cell

Identify the two relevant half-reactions and the polarity of each electrode.

The more positive of the two systems is the positive terminal of the cell.
$Ag^+(aq) + e^- \rightleftharpoons Ag(s)$ $E^\ominus = +0.80$ V *positive terminal*
$Fe^{2+}(aq) + 2e^- \rightleftharpoons Fe(s)$ $E^\ominus = -0.44$ V *negative terminal*

The standard electrode potential is the difference between the $E^\ominus$ values:

$E^\ominus_{cell}$ is the difference between the standard electrode potentials.

Simply subtract the $E^\ominus$ of the negative terminal from the $E^\ominus$ of the positive terminal:
$$E^\ominus_{cell} = E^\ominus \text{ (positive terminal)} - E^\ominus \text{ (negative terminal)}$$
$$\therefore E^\ominus_{cell} = 0.080 - (-0.44) = \mathbf{1.24 \text{ V}}$$

Determination of the cell reaction in a silver-iron(II) cell

Work out the actual direction of the half-equations.

The more negative half-equation is reversed. The negative terminal is then on the left-hand side.

The **half-equation** with the **more negative $E^\ominus$** value provides the electrons and proceeds to the left. This half-equation is reversed so that the two half-reactions taking place can be clearly seen and compared.

$Ag^+(aq) + e^- \longrightarrow Ag(s)$ *reaction at positive terminal*
$Fe(s) \longrightarrow Fe(aq) + 2e^-$ *reaction at negative terminal*

The overall cell reaction can be found by adding the half-equations.

The silver half-equation must first be multiplied by '2' to balance the electrons.
The two half-equations are then added.

$2Ag^+(aq) + Fe(s) \longrightarrow 2Ag(s) + Fe^{2+}(aq)$

Cell diagrams

AQA M5

A simple metal electrode comprises a metal in contact with a solution of its ions, e.g. $Cu(s)|Cu^{2+}(aq)$. This is the half-cell.

- The solid vertical line shows a boundary between a solid and a solution.

A cell diagram is a convenient way of representing a cell.

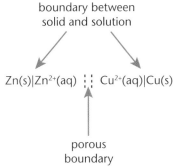

Two dotted vertical lines show a porous boundary between two half-cells. This is the salt bridge in simple cells.

Notice that the ions in solution are shown in contact with the salt bridge. The metals taking the electric current in and out of the cell are shown at either end of the cell diagram.

The standard cell potential of the cell, as shown in the cell diagram, is defined as:
$$E^\ominus(\text{cell}) = E^\ominus(\text{right-hand side}) - E^\ominus(\text{left-hand side})$$

The sign of the cell gives the polarity of the right-hand half-cell.

For the cell shown above,
- $E^\ominus(\text{cell}) = 0.34 - (-0.76) = +1.10$ V
- the $Cu(s)|Cu^{2+}(aq)$ half-cell (shown on the right) is positive.

51

Energy changes in chemistry

Electrochemical cells

Electrochemical cells can be used as a commercial source of electrical energy.

Cells can be divided into three main types:

- non-rechargeable cells – the cell is used until the chemicals are used up to such a point at which the voltage falls
- rechargeable cells – the cell reaction can readily be reversed when recharging, allowing the cell to be used again
- fuel cells – the cell reaction requires external supplies of a fuel and an oxidant, which are consumed and need to be replenished.

An electrochemical cell comprises two half-cells with different electrode potentials. For example, a simple cell can be set up based on zinc and copper.

The redox systems are shown below.

$Zn^{2+}(aq) + 2e^- \rightleftharpoons Zn(s)$ $\quad\quad E^\ominus = -0.74$ V

$Cu^{2+}(aq) + 2e^- \rightleftharpoons Cu(s)$ $\quad\quad E^\ominus = +0.34$ V

The more negative zinc system provides the electrons:

Overall: $Zn(s) + Cu^{2+}(aq) \longrightarrow Cu(s) + Zn^{2+}(aq)$

Chargeable and non-rechargeable cells are used as our modern-day cells and batteries.

Other common examples include:
- nickel and cadmium (NiCad) batteries used in rechargeable batteries
- lithium 'button' cells used in watches
- lithium-ion batteries used in laptops.

All these cells work on the same principle, using two redox systems.

Fuel cells for energy

In a fuel cell, energy from the reaction of a fuel with oxygen is used to create a voltage.

- The reactants flow in and products flow out, while the electrolyte remains in the cell.
- Fuel cells can operate virtually continuously as long as the fuel and oxygen continue to flow into the cell. Fuel cells do not have to be recharged.

In the future, a 'hydrogen economy' based on hydrogen fuel cells may well contribute greatly to energy needs.

The diagram below shows a simple hydrogen–oxygen fuel cell.

- At the negative electrode (cathode),

    $H_2(g) \longrightarrow 2H^+(aq) + 2e^-$

- At the positive electrode (anode),

    $\frac{1}{2}O_2(g) + 2H^+(aq) + 2e^- \longrightarrow H_2O(l)$

- Overall: $\quad H_2(g) + \frac{1}{2}O_2(g) \longrightarrow H_2O(l)$

In exams, you are unlikely to have to recall a specific storage cell.

You might be expected to make predictions on a supplied electrochemical cell. All relevant electrode potentials and other data would be supplied though.

Other hydrogen fuel cells operate with an alkaline electrolyte.

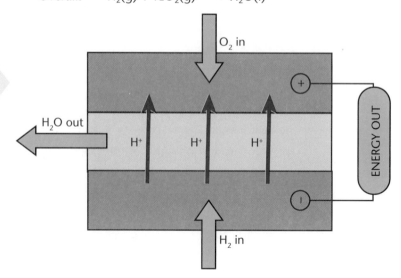

Energy changes in chemistry

Development of fuel cell vehicles (FCVs)

Fuel cells are being developed as an alternative to finite oil-based fuels in cars.

Scientists in the car industry are developing fuel cell vehicles (FCVs), fuelled by:

- hydrogen gas
- hydrogen-rich fuels.

Hydrogen-rich fuels include methanol, natural gas, or petrol. These are converted into hydrogen gas by an onboard 'reformer'. These fuels are easier to transport and deliver into an FCV than hydrogen gas.

FCVs have advantages over conventional petrol or diesel-powered vehicles.

- Pure hydrogen emits only water.
- Hydrogen-rich fuels produce only small amounts of air pollutants and CO_2.
- Efficiency can be more than twice that of similarly-sized conventional vehicles. Other advanced technologies are being investigated to further increase efficiency.

Hydrogen might be stored in FCVs:

- as a liquid under pressure
- adsorbed on the surface of a solid material
- absorbed within a solid material.

> Advantages of hydrogen fuel cells:
> less pollution and less CO_2,
> greater efficiency.

Limitations of hydrogen fuel cells

There are logistical problems in the development of hydrogen fuel cells:

- storing and transporting hydrogen, in terms of safety, feasibility of a pressurised liquid and a limited life cycle of a solid 'adsorber' or 'absorber'
- limited lifetime (requiring regular replacement and disposal) and high production costs
- use of toxic chemicals in their production
- initial manufacture of hydrogen requires energy
- public and political acceptance of hydrogen as a fuel.

> Liquefying hydrogen is costly in terms of energy use.
> Safety is a major concern: Hydrogen is very flammable.
>
> Hydrogen is an 'energy carrier' and not an 'energy source'. Hydrogen is first made either by electrolysis of water or by reacting methanol (a finite fuel) with steam.

Future energy needs

Political and social desire to move to a hydrogen economy has many obstacles. Many people do not realise that energy is needed to produce hydrogen and that fuel cells have a finite life.

Progress check

1. Use the standard electrode potentials on page 50, to answer the questions that follow.
   (a) Calculate the standard cell potential for the following cells.
      (i) F_2/F^- and Ag^+/Ag
      (ii) Ag^+/Ag and Cr^{3+}/Cr
      (iii) Cu^{2+}/Cu and Cr^{3+}/Cr.
   (b) For each cell in (a) determine which half-reactions proceed, and hence construct, the cell reaction.

1 (a) (i) 2.07 V (ii) 1.57 V (iii) 1.11 V.
(b) $F_2 + 2Ag \longrightarrow 2F^- + 2Ag^+$
$3Ag^+ + Cr \longrightarrow 3Ag + Cr^{3+}$
$3Cu^{2+} + 2Cr \longrightarrow 3Cu + 2Cr^{3+}$

Energy changes in chemistry

2.4 Predicting redox reactions

After studying this section you should be able to:

- use standard cell potentials to predict the feasibility of a reaction
- recognise the limitations of such predictions

LEARNING SUMMARY

Key points from AS

- **Redox reactions**
  Revise AS pages 33–35

During AS Chemistry, you learnt how to combine two half-equations to construct a full redox equation.

Standard electrode potentials can be used to predict the direction of any aqueous redox reaction from the relevant half-reactions.

Using standard electrode potentials to make predictions

By studying standard electrode potentials, predictions can be made about the likely feasibility of a reaction.

The table below shows three redox systems.

	redox system	$E^\ominus$ /V
A	$H_2O_2(aq) + 2H^+ + 2e^- \rightleftharpoons 2H_2O(l)$	+1.77
B	$I_2(aq) + 2e^- \rightleftharpoons 2I^-(aq)$	+0.54
C	$Fe^{3+}(aq) + 3e^- \rightleftharpoons Fe(s)$	–0.04

We can predict the redox reactions that may take place between these species by treating the redox systems as if they are half-cells.

Comparing the redox systems A and B

- The I_2/I^- system has the less positive $E^\ominus$ value and its half-reaction will proceed to the left, donating electrons:

$$H_2O_2(aq) + 2H^+(aq) + 2e^- \longrightarrow 2H_2O(l) \qquad E^\ominus = +1.77 \text{ V}$$
$$I_2(aq) + 2e^- \longleftarrow 2I^-(aq) \qquad E^\ominus = +0.54 \text{ V}$$

We would therefore predict that **H_2O_2 and H^+** would react with **I^-**.

Comparing the redox systems B and C

- The Fe^{3+}/Fe system has the more negative $E^\ominus$ value and its half-reaction will proceed to the left, donating electrons:

$$I_2(aq) + 2e^- \longrightarrow 2I^-(aq) \qquad E^\ominus = +0.54 \text{ V}$$
$$Fe^{3+}(aq) + 3e^- \longleftarrow Fe(s) \qquad E^\ominus = -0.04 \text{ V}$$

We would therefore predict that **I_2** would react with **Fe**.

Comparing the redox systems A and C

- The Fe^{3+}/Fe system has the more negative $E^\ominus$ value and its half-reaction will proceed to the left, donating electrons:

$$H_2O_2(aq) + 2H^+(aq) + 2e^- \longrightarrow 2H_2O(l) \qquad E^\ominus = +1.77 \text{ V}$$
$$Fe^{3+}(aq) + 3e^- \longleftarrow Fe(s) \qquad E^\ominus = -0.04 \text{ V}$$

We would therefore predict that **H_2O_2 and H^+** would react with **Fe**.

Notice that a reaction takes place between reactants from different sides of each half equation.

> **KEY POINT**
>
> When comparing two redox systems, we can predict that a reaction may take place between:
> - the stronger reducing agent with the more negative $E^\ominus$, on the right-hand side of the redox system
> - the stronger oxidising agent with the more positive $E^\ominus$, on the left-hand side of the redox system.

Energy changes in chemistry

Limitations of standard electrode potentials

AQA M5

Electrode potentials and ionic concentration

Non-standard conditions alter the value of an electrode potential.

The half-equation for the copper half-cell is shown below.

$$Cu^{2+}(aq) + 2e^- \rightleftharpoons Cu(s)$$

Using le Chatelier's principle, by increasing the concentration of $Cu^{2+}(aq)$:

- the equilibrium opposes the change by moving to the right
- electrons are removed from the equilibrium system
- the electrode potential becomes more positive.

> A change in electrode potential resulting from concentration changes means that predictions made on the basis of the **standard** value may not be valid.

Will a reaction actually take place?

Remember that these are equilibrium processes.

- Predictions can be made but these give no indication of the reaction rate which may be extremely slow, caused by a large activation energy.
- The actual conditions used may be different from the standard conditions used to record $E^\ominus$ values. This will affect the value of the electrode potential (see above).
- Standard electrode potentials apply to aqueous equilibria. Many reactions take place that are not aqueous.

As a general working rule:

- the larger the difference between $E^\ominus$ values, the more likely that a reaction will take place
- if the difference between $E^\ominus$ values is less than 0.4 V, then a reaction is unlikely to go to completion.

Progress check

1. For each of the three predicted reactions on page 54, construct full redox equations.
2. Use the standard electrode potentials below to answer the questions that follow.

   $Fe^{3+}(aq) + e^- \rightleftharpoons Fe^{2+}(aq)$ $E^\ominus = +0.77$ V
   $Br_2(l) + 2e^- \rightleftharpoons 2Br^-(aq)$ $E^\ominus = +1.07$ V
   $Ni^{2+}(aq) + 2e^- \rightleftharpoons Ni(s)$ $E^\ominus = -0.25$ V
   $O_2(g) + 4H^+(aq) + 4e^- \rightleftharpoons 2H_2O(l)$ $E^\ominus = +1.23$ V

   (a) Arrange the reducing agents in order with the most powerful first.
   (b) Predict the reactions that could take place.

1. A and B: $H_2O_2(aq) + 2H^+(aq) + 2I^-(aq) \longrightarrow I_2(aq) + 2H_2O(l)$; B and C: $2Fe(s) + 3I_2(aq) \longrightarrow 2Fe^{3+}(aq) + 6I^-(aq)$
   A and C: $3H_2O_2(aq) + 6H^+(aq) + 2Fe(s) \longrightarrow 2Fe^{3+}(aq) + 6H_2O(l)$
2. a) Ni(s), $Fe^{2+}(aq)$, $Br^-(l)$, $H_2O(l)$.
   b) Br_2 with Fe^{2+}; Fe^{3+} with Ni; O_2 and H^+ with Fe^{2+}; Br_2 with Ni; O_2 and H^+ with Br^-; O_2 and H^+ with Ni.

Energy changes in chemistry

Sample question and model answer

The standard electrode potentials of three half reactions are given below:

$$MnO_4^-(aq) + 8H^+(aq) + 5e^- \rightleftharpoons Mn^{2+}(aq) + 4H_2O(aq) \quad E^\ominus = +1.52 \text{ V}$$
$$Cl_2(g) + 2e^- \rightleftharpoons 2Cl^-(aq) \quad E^\ominus = +1.36 \text{ V}$$
$$Fe^{3+}(aq) + e^- \rightleftharpoons Fe^{2+}(aq) \quad E^\ominus = +0.77 \text{ V}$$

(a) $Cl_2(g)$ is produced in a redox process.

> To form Cl_2:
> $Cl_2 + 2e^- \leftarrow 2Cl^-$
> This reaction has been reversed and must be a negative electrode.
> It can react with a half equation in the opposite direction with a more positive electrode potential:
> $MnO_4^- + 8H^+ + 5e^- \rightarrow$

(i) Justify whether the conversion of $Cl^-(aq)$ to $Cl_2(g)$ is oxidation or reduction.

Oxidation because electrons are lost ✓ during the conversion.

(ii) Suggest what ions could be used for this conversion.

$MnO_4^-(aq)/H^+(aq)$ ✓

> 'Chemicals' or 'reagents' refer to actual materials that you would use in the laboratory to carry out a reaction.
> Here, $KMnO_4$ and HCl are the reagents although only MnO_4^- and H^+ are involved in the reaction

(iii) State the conditions needed.

acidified ✓

(iv) What chemicals could be used to produce $Cl_2(g)$?

$KMnO_4$ ✓ and concentrated HCl ✓ [5]

(b) (i) Calculate the standard cell potential for the reaction between $Fe^{2+}(aq)$ and acidified $MnO_4^-(aq)$.

$E^\ominus_{cell} = E^\ominus(+ \text{ terminal}) - E^\ominus(- \text{ terminal})$

∴ $E^\ominus_{cell} = +1.52 - (+0.77) = 0.75 \text{ V}$ ✓

> Simply subtract the $E^\ominus$ of the negative terminal from the $E^\ominus$ of the positive terminal.

(ii) Construct a balanced equation for this reaction.

$MnO_4^-(aq) + 8H^+(aq) + 5e^- \rightarrow Mn^{2+}(aq) + 4H_2O(l)$

$Fe^{2+}(aq) \rightarrow Fe^{3+}(aq) + e^-$

> Reverse the more negative half-reaction to give each half-cell reactions.

Balance electrons: $5Fe^{2+}(aq) \rightarrow 5Fe^{3+}(aq) + 5e^-$

$5Fe^{2+}(aq) + MnO_4^-(aq) + 8H^+(aq) \rightarrow 5Fe^{3+}(aq) + Mn^{2+}(aq) + 4H_2O(l)$

> The overall cell reaction can be found by adding the balanced half equations.

use of (× 5) to balance electrons ✓ balanced equation ✓ [3]

(c) Suggest why hydrochloric acid is not used to acidify the $MnO_4^-(aq)$ in the reaction in (b)(i).

MnO_4^- would react with the Cl^- ✓ present in hydrochloric acid. [1]

[Total: 9]

Energy changes in chemistry

Practice examination questions

1 The Born–Haber cycle for the formation of potassium bromide from potassium and bromine may be represented by a series of stages labelled **A** to **F** as shown.

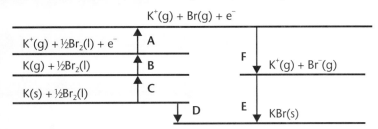

(a) (i) Write the letters **A** to **F** next to the corresponding definition in the table below.

definition	letter	ΔH/kJ mol^{-1}
1st ionisation energy of potassium		+419
1st electron affinity of bromine		−325
the enthalpy of atomisation of potassium		+89
the enthalpy of atomisation of bromine		+112
the lattice enthalpy of potassium bromide		
the enthalpy of formation of potassium bromide		−394

(ii) Calculate the lattice enthalpy of potassium bromide from the data given. [5]

(b) Explain whether you would expect the lattice enthalpies of the following compounds to be more or less exothermic than that of potassium bromide.

(i) KI; (ii) CaBr$_2$ [4]

[Total: 9]

2 (a) Write equations, including state symbols, for the following:

(i) the lattice enthalpy of magnesium chloride

(ii) the enthalpy change of hydration of a Mg^{2+} ion

(iii) the enthalpy change of solution of magnesium chloride. [3]

(b) The lattice enthalpy of MgCl$_2$ is −2526 kJ mol^{-1}.

The enthalpy change of hydration of Mg^{2+} is −1891 kJ mol−1.

The enthalpy change of hydration of Cl$^-$ is −384 kJ mol−1.

Calculate the enthalpy change of solution of magnesium chloride. [3]

[Total: 6]

3 Chromium can be prepared by the reduction of chromium(III) oxide with aluminium.

$$Cr_2O_3(s) + 2Al(s) \longrightarrow 2Cr(s) + Al_2O_3(s)$$

The table below gives information about the reactants and products.

substance	Cr$_2$O$_3$(s)	Al(s)	Cr(s)	Al$_2$O$_3$(s)
ΔH_f/kJ mol^{-1}	−602	0	0	−1676
S/J K^{-1} mol^{-1}	27	28	81	51

Calculate the free energy change for this reaction at 298 K and comment on its feasibility at this temperature.

[Total: 8]

Energy changes in chemistry

Practice examination questions (continued)

4 A standard cell was set up as shown below.

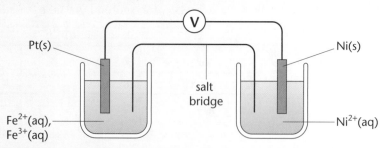

(a) What conditions are necessary for this cell to be a standard cell? [2]

(b) What is the purpose of the salt bridge? [1]

(c) The standard electrode potentials for the half-cells above are:

$$Ni^{2+} + 2e^- \rightleftharpoons Ni \qquad E^\ominus = -0.25 \text{ V}$$
$$Fe^{3+} + e^- \rightleftharpoons Fe^{2+} \qquad E^\ominus = +0.77 \text{ V}$$

(i) Calculate the standard cell potential of the complete cell.

(ii) Draw an arrow on the diagram above to show the direction of electron flow in the external circuit.

(iii) What is the polarity of each electrode? [3]

(d) (i) Write equations for the reaction at each electrode.

(ii) Hence write an equation to show the overall reaction. [3]

(e) The solution in the nickel half-cell was replaced by a solution that had a higher concentration of Ni^{2+}(aq) ions. Predict the effect on the cell potential. Explain your reasoning. [2]

[Total: 11]

5 (a) Use the standard electrode potentials given below to answer the questions that follow.

			$E^\ominus$/ V
Mg^{2+}(aq) + 2e$^-$	$\rightleftharpoons$	Mg(s)	−2.37
I_2(aq) + 2e$^-$	$\rightleftharpoons$	2I$^-$(aq)	+0.54
Mn^{3+}(aq) + e$^-$	$\rightleftharpoons$	Mn^{2+}(aq)	+1.49

(i) Which of the six species above is the strongest oxidising agent?

(ii) Predict, with an equation and a reason, which species above would react with I$^-$(aq) ions to form iodine, I_2(aq). [4]

(b) The standard electrode potentials for two redox equilibria are given below.

$$O_2(g) + 2H_2O(l) + 4e^- \rightleftharpoons 4OH^-(aq) \qquad E^\ominus = +0.40 \text{ V}$$
$$V^{3+}(aq) + e^- \rightleftharpoons V^{2+}(aq) \qquad E^\ominus = -0.26 \text{ V}$$

Two half-cells based on these redox equilibria are connected to make a cell. Determine the cell potential, $E^\ominus$, for this cell and write an equation for the cell reaction. [3]

[Total: 7]

Chapter 3
The Periodic Table

The following topics are covered in this chapter:

- Oxides and chlorides of the Period 3 elements
- Chemistry of the p-block elements
- The transition elements
- Transition element complexes
- Ligand substitution of complex ions
- Redox reactions of transition metal ions
- Catalysis
- Reactions of metal aqua-ions

Key points from AS

- Bonding and structure
  Revise AS pages 39–55

During AS Chemistry, you learnt about physical trends in properties of **elements** across Period 3. Across Period 3, **compounds** of the elements also show characteristic trends. For A2 Chemistry, physical and chemical trends in the properties of Period 3 oxides and chlorides are explained in terms of bonding and structure.

3.1 Chemistry of the Period 3 elements

After studying this section you should be able to:

- describe reactions of Period 3 elements with water
- describe and explain the periodic variation of the formulae and boiling points of the main oxides of Period 3
- describe the reactions of Period 3 elements with oxygen
- describe the action of water on Period 3 oxides
- describe these reactions in terms of structure and bonding
- describe the reaction of Period 3 oxides with acids and bases

The action of water on the Period 3 elements

Reactions of sodium and magnesium

Sodium and magnesium metals both react with water forming an alkaline solution and hydrogen gas. The reaction with sodium is extremely reactive but magnesium reacts slowly.

$$2Na(s) + 2H_2O(l) \longrightarrow 2NaOH(aq) + H_2(g) \quad pH = 14$$
$$Mg(s) + 2H_2O(l) \longrightarrow Mg(OH)_2(aq) + H_2(g) \quad pH = 11$$

Each hydroxide dissociates in water, releasing OH^- ions into solution:

$$NaOH(aq) \longrightarrow Na^+(aq) + OH^-(aq)$$
$$Mg(OH)_2(aq) \longrightarrow Mg^{2+}(aq) + 2OH^-(aq)$$

Preparation of oxides from the elements

Oxides of the Period 3 elements Na to Cl can be prepared by heating each element in oxygen. These are redox reactions:

- the Period 3 element is oxidised
- oxygen is reduced.

Metal oxides

Equations for the reactions of oxygen with the metals sodium, magnesium and aluminium are shown below.

At room temperature and pressure:
Na_2O, MgO and Al_2O_3 are white solids.

$$4Na(s) + O_2(g) \xrightarrow{\text{yellow flame}} 2Na_2O(s)$$
$$2Mg(s) + O_2(g) \xrightarrow{\text{white flame}} 2MgO(s)$$
$$4Al(s) + 3O_2(g) \xrightarrow{\text{white flame}} 2Al_2O_3(s)$$

The Periodic Table

Phosphorus and sulfur exist as P_4 and S_8 molecules.

Sulfur, S_8, is usually simplified in equations as S.

At room temperature and pressure:
- P_4O_{10} is a white solid
- SO_2 is a colourless gas
- SO_3 is a colourless liquid.

- Na_2O and MgO are **ionic compounds**.
- Al_2O_3 has bonding that is **intermediate** between ionic and covalent.

Non-metal oxides

Equations for the reactions of phosphorus and sulfur with oxygen are shown below.

$$P_4(s) + 5O_2(g) \xrightarrow{\text{white flame}} P_4O_{10}(s)$$

$$S(s) + O_2(g) \xrightarrow{\text{blue flame}} SO_2(g)$$

SO_2 reacts further with oxygen in the presence of a catalyst (V_2O_5):

$$2SO_2(g) + O_2(g) \longrightarrow 2SO_3(l)$$

- P_4O_{10}, SO_2 and SO_3 are **covalent compounds**.

The formulae of the Period 3 oxides

AQA M5

Key points from AS
- The modern Periodic Table
  Revise AS pages 56–59
- Redox reactions
  Revise AS pages 33–35

The highest oxidation state of an element is usually the group number.

Chlorine forms several oxides. The highest is Cl_2O_7.

Across Period 3, the formulae of compounds show a regular pattern which depends upon the **number of electrons** in the outer shell.

The **oxidation state** of an element indicates the **number of electrons** involved in bonding. The relationship of the formulae of the Period 3 oxides with electronic structure and oxidation state is shown below.

element	Na	Mg	Al	Si	P	S	Cl
electrons in outer shell	1	2	3	4	5	6	7
formula of oxide	Na_2O	MgO	Al_2O_3	SiO_2	P_4O_6 P_4O_{10}	SO_2 SO_3	Cl_2O_7
oxidation state of element in oxide	+1	+2	+3	+4	+3 +5	+4 +6	+7

→ increase in highest oxidation state

The amount of oxygen atoms per mole of the element increases steadily across Period 3.

For each element in the Period, there is an increase of 0.5 mole of O per mole of element.

$Na_1O_{0.5}$ Mg_1O_1 $Al_1O_{1.5}$ Si_1O_2 $P_1O_{2.5}$ S_1O_3 $Cl_1O_{3.5}$

Periodicity means a repeating trend. This is one of the important ideas of chemistry – a trend across one period is repeated across other periods.

This periodicity is also seen across Periods 2 and 4:

Period 2	Li_2O	BeO	B_2O_3	CO_2
Period 3	Na_2O	MgO	Al_2O_3	SiO_2
Period 4	K_2O	CaO	Ga_2O_3	GeO_2

Trend in the boiling points of the Period 3 oxides

AQA M5

Key points from AS
- Bonding, structure and properties
  Revise AS pages 52–55
- Trends in boiling points of Period 3 elements
  Revise AS pages 58–59

The table below compares the boiling points of some Period 3 oxides with their structure and bonding.

oxide	Na_2O	MgO	Al_2O_3	SiO_2	P_4O_{10}	SO_3	Cl_2O_7
boiling point /°C	1275	3600	2980	2230	605	45	83
structure	← giant lattice →					simple molecules	
bonding	← ionic →		mixed	covalent		covalent and van der Waals'	
	strong forces between ions			strong forces between atoms		weak van der Waals' forces between molecules	

The Periodic Table

During your study of AS Chemistry, you learnt that the boiling points of the elements and compounds are dependent upon bonding and structure.

The same principles can be applied many times in chemistry.

Giant structures and boiling point

The oxides with **giant** structures ($Na_2O \rightarrow SiO_2$):
- have **high boiling points**
- require a large amount of energy to break the **strong forces** between their particles.

The particles making up a giant lattice can be different.
- Na_2O and MgO have ionic bonding and the **strong** forces are electrostatic attractions acting between positive and negative **ions**.
- Al_2O_3 has bonding intermediate between ionic and covalent. **Strong** forces act between particles that are **intermediate** between ions and atoms.
- SiO_2 has covalent bonding and the **strong** forces are 'shared pairs of electrons' acting between the **atoms**.

P_4O_{10}, SO_2, SO_3 and Cl_2O_7 all exist as covalently bonded molecules. The forces between the molecules though are van der Waals' forces.

Simple molecular structures and boiling point

The oxides with **simple molecular** structures ($P_4O_{10} \rightarrow Cl_2O_7$):
- have low boiling points
- require a small amount of energy to break the **weak van der Waals' forces** acting between the **molecules**.

The action of water on the Period 3 oxides

AQA   M5

Sodium and magnesium metals react with water forming the metal hydroxide and hydrogen gas. The reaction with sodium is rapid but magnesium reacts very slowly.

Key points from AS
- **The Group 1 elements**
  Revise AS pages 60–61
- **The Group 2 elements**
  Revise AS pages 61–62

Reactions of sodium and magnesium oxides

Metal oxides form alkalis in water.

$Na_2O(s) + 2H_2O(l) \longrightarrow 2NaOH(aq)$ pH = 14

$MgO(s) + 2H_2O(l) \longrightarrow Mg(OH)_2(aq)$ pH = 11

Although MgO reacts with water, the magnesium hydroxide formed is only slightly soluble in water forming a weak alkaline solution only.

Each hydroxide dissociates in water, releasing OH^- ions into solution:

$NaOH(aq) \longrightarrow Na^+(aq) + OH^-(aq)$

Oxides of aluminium and silicon

Al_2O_3 and SiO_2 have very strong lattices which cannot be broken down by water. Consequently these compounds are **insoluble** in water.

Reactions of non-metal oxides

Non-metal oxides form acids in water. Equations for the reactions of $P_4O_{10}(s)$, $SO_2(g)$ and $SO_3(l)$ with water are shown below.

$P_4O_{10}(s) + 6H_2O(l) \longrightarrow 4H_3PO_4(aq)$ *phosphoric acid*

$SO_2(g) + H_2O(l) \longrightarrow H_2SO_3(aq)$ *sulfurous acid*

$SO_3(l) + H_2O(l) \longrightarrow H_2SO_4(aq)$ *sulfuric acid*

$Cl_2O_7(s) + H_2O(l) \longrightarrow 2HClO_4(aq)$ *chloric(VII) acid*

Each acid dissociates in the water, releasing H^+ ions into solution.

E.g. $H_2SO_4(aq) \longrightarrow H^+(aq) + HSO_4^-(aq)$

> **KEY POINT**
> An important general rule is:
> - **metal oxides** are **basic**
>   When soluble, metal oxides form alkaline solutions in water.
> - **non-metal oxides** are **acidic**
>   When soluble, non-metal oxides form acidic solutions in water.

The Periodic Table

The action of water on Period 3 oxides is summarised below.

The reactions of oxides with acids and bases

Basic oxides, Na_2O and MgO

Basic oxides are **neutralised** by acids to form a salt and water.

e.g. $Na_2O(s) + 2HCl(aq) \longrightarrow 2NaCl(aq) + H_2O(l)$

$MgO(s) + H_2SO_4(aq) \longrightarrow MgSO_4(aq) + H_2O(l)$

Acidic oxides, P_4O_{10}, SO_2, SO_3 and Cl_2O_7

Acidic oxides are **neutralised** by bases to form a salt and water.

e.g. $P_4O_{10}(s) + 12NaOH(aq) \longrightarrow 4Na_3PO_4(aq) + 6H_2O(l)$

$SO_2(g) + 2KOH(aq) \longrightarrow K_2SO_3(aq) + H_2O(l)$

$SO_3(s) + 2KOH(aq) \longrightarrow K_2SO_4(aq) + H_2O(l)$

$Cl_2O_7(s) + 2NaOH(aq) \longrightarrow 2NaClO_4(aq) + H_2O(l)$

Aluminium oxide, Al_2O_3

Al_2O_3 is **neutralised** by both acids and bases and is **amphoteric**.

> An **amphoteric** oxide is one that is neutralised by both acids and bases.

With hot dilute $HCl(aq)$,

$Al_2O_3(s) + 6HCl(aq) \longrightarrow 2AlCl_3(aq) + 3H_2O(l)$

With hot concentrated $NaOH(aq)$,

$Al_2O_3(s) + 2NaOH(aq) + 3H_2O(l) \longrightarrow 2NaAl(OH)_4(aq)$

Silicon(IV) oxide, SiO_2

SiO_2 is **neutralised** by hot concentrated bases:

$2KOH(aq) + SiO_2(s) \longrightarrow K_2SiO_3(aq) + H_2O(l)$

Progress check

1. (a) Across the Periodic Table, what pattern is shown in the formulae of the Period 3 oxides?
   (b) Why do elements in the **same** group form compounds with **similar** formulae?
2. The reactions of elements with oxygen are redox reactions. Use oxidation numbers to show that the reaction of aluminium with oxygen is a redox reaction.

1. (a) For each element in the period, there is an increase of 0.5 mole of O per mole of the element.
   (b) Although the outer shell is different, there is the same number of electrons in the outer shell.
2. $4Al(s) + 3O_2(g) \longrightarrow 2Al_2O_3(s)$
   Al changes oxidation number from 0 to +3 (oxidation)
   O changes oxidation state from 0 to −2 (reduction)

The Periodic Table

3.2 The transition elements

After studying this section you should be able to:

- explain what is meant by the terms d-block element and transition element
- deduce the electronic configurations of atoms and ions of the d-block elements Sc → Zn
- know the typical properties of a transition element

Key points from AS

- The Periodic Table
  Revise AS page 56–66

During AS Chemistry, you studied the metals in the s-block of the Periodic Table. For A2 Chemistry, this understanding is extended with a detailed study of the elements scandium to zinc in the d-block of the Periodic Table. Many of the principles introduced during AS Chemistry will be revisited, especially redox chemistry, and structure and bonding.

The d-block elements

AQA M5

The d-block is found in the centre of the Periodic Table. For A2 Chemistry, you will be studying mainly the d-block elements in Period 4 of the Periodic Table: Sc, Ti, V, Cr, Mn, Fe, Co, Ni, Cu and Zn.

Electronic configurations of d-block elements

The diagram below shows the sub-shell being filled across Period 4 of the Periodic Table.

4s		3d									4p					
K	Ca	Sc	Ti	V	Cr	Mn	Fe	Co	Ni	Cu	Zn	Ga	Ge	As	Se	Br	Kr

Key points from AS

- Sub-shells and orbitals
  Revise AS pages 15–17
- Filling the sub-shells
  Revise AS pages 17–18
- Sub-shells and the Periodic Table
  Revise AS page 18

- The elements Sc ⟶ Zn have their highest energy electrons in the 4s and 3d sub-shells.
- Across the d-block, electrons are filling the 3d sub-shell. The diagram below shows the outermost electron shell for Sc ⟶ Zn.

Notice how the orbitals are filled:

- the orbitals in the 3d sub-shell are first occupied singly to prevent any repulsion caused by pairing.

*The 3d sub-shell is at a higher energy and fills **after** the 4s sub-shell.*

Notice chromium and copper:

- chromium has one electron in each orbital of the 4s and 3d sub-shells ⟶ extra stability

See also the stability from a half-full p sub-shell.

- copper has a full 3d sub-shell ⟶ extra stability.

		4s	3d				
Sc	[Ar] $3d^1 4s^2$	↑↓	↑				
Ti	[Ar] $3d^2 4s^2$	↑↓	↑	↑			
V	[Ar] $3d^3 4s^2$	↑↓	↑	↑	↑		
Cr	[Ar] $3d^5 4s^1$	↑	↑	↑	↑	↑	↑
Mn	[Ar] $3d^5 4s^2$	↑↓	↑	↑	↑	↑	↑
Fe	[Ar] $3d^6 4s^2$	↑↓	↑↓	↑	↑	↑	↑
Co	[Ar] $3d^7 4s^2$	↑↓	↑↓	↑↓	↑	↑	↑
Ni	[Ar] $3d^8 4s^2$	↑↓	↑↓	↑↓	↑↓	↑	↑
Cu	[Ar] $3d^{10} 4s^1$	↑	↑↓	↑↓	↑↓	↑↓	↑↓
Zn	[Ar] $3d^{10} 4s^2$	↑↓	↑↓	↑↓	↑↓	↑↓	↑↓

Transition elements

Successive ionisation energies provide evidence for electronic configurations of the d-block elements Sc to Zn. This is most evident by the presence of peaks matching the $3d^5$ half-filled shell.

The elements in the d-block, Ti→Cu, are called transition elements.

> **KEY POINT**
> - A transition element has **at least one ion** with a **partially-filled d sub-shell**.

Although scandium and zinc are d-block elements, they are **not** *transition elements*. Neither element forms an ion with a partially filled d sub-shell.

The Periodic Table

Scandium forms one ion only, Sc^{3+} with the 3d sub-shell **empty**.

$$Sc\ [Ar]3d^1 4s^2 \xrightarrow{\text{loss of 3 electrons}} Sc^{3+} \text{ only: } [Ar]$$

Zinc forms one ion only, Zn^{2+} with the 3d sub-shell **full**.

$$Zn\ [Ar]3d^{10} 4s^2 \xrightarrow{\text{loss of 2 electrons}} Zn^{2+} \text{ only: } [Ar]3d^{10}$$

d–block elements: Sc | Ti | V | Cr | Mn | Fe | Co | Ni | Cu | Zn

transition elements

Key points from AS
- **The s-block elements**
  Revise AS pages 60–62

Properties of transition elements

1. high density
2. high m. pt. and b. pt.
3. moderate to low reactivity
4. two or more oxidation states
5. coloured ions
6. complex ions with ligands
7. catalysts

Typical properties of transition elements

The transition elements are all metals – they are good conductors of heat and electricity. In addition, some general properties distinguish the transition elements from the s-block metals.

Density
- The transition elements are more **dense** than other metals.
  - They have smaller atoms than the metals in Groups 1 and 2. The small atoms are able to pack closely together ⟶ high density.

Melting point and boiling point
- The transition elements have **higher melting and boiling points** than other metals.
  - Within the metallic lattice, ions are **smaller** than those of the s-block metals – the metallic bonding between the metal ions and the delocalised electrons is strong.

Reactivity
- The transition elements have **moderate to low reactivity**.
  - Unlike the s-block metals, they do **not** react with cold water. Many transition metals react with dilute acids although some, such as gold, silver and platinum, are extremely unreactive.

Oxidation states
- Most transition elements have compounds with **two or more oxidation states**.

Colour of compounds
- The transition elements have **coloured compounds and ions**.
  - Most transition metals have at least one oxidation state that is coloured.

Complex ions
- The ions of transition elements form **complex ions** with ligands.

Catalysis
- Many transition elements can act as **catalysts**.

Formation of ions

The energy levels of the 4s and 3d sub-shells are very close together.

> **KEY POINT**
> - Transition elements are able to form **more than one ion**, each with a different oxidation state, by losing the **4s** electrons and different numbers of **3d** electrons.
> - When forming ions, the 4s electrons are **lost first**, before the 3d electrons.

The Periodic Table

The Fe^{3+} ion, $[Ar]3d^5$ is more stable than the Fe^{2+} ion, $[Ar]3d^54s^1$. Fe^{3+} has a half-full 3d sub-shell ⟶ stability.

The 4s and 3d energy levels are very close together and both are involved in bonding.

When forming ions, the 4s electrons are **lost before** the 3d electrons. Note that this is different from the order of **filling** – 4s before 3d.

Once the 3d sub-shell starts to fill, the 4s electrons are repelled to a higher energy level and are lost first.

Forming Fe^{2+} and Fe^{3+} ions from iron

Iron has the electronic configuration $[Ar]3d^64s^2$

Iron forms two common ions, Fe^{2+} and Fe^{3+}.

An Fe atom loses **two 4s electrons** ⟶ Fe^{2+} ion.

An Fe atom loses **two 4s electrons and one 3d electron** ⟶ Fe^{3+} ion.

The variety of oxidation states

The table below summarises the stable oxidation states of the elements scandium to zinc. Those in bold type represent the commonest oxidation numbers of the elements in their compounds.

Sc	Ti	V	Cr	Mn	Fe	Co	Ni	Cu	Zn
	+1	+1	+1	+1	+1	+1	+1	**+1**	
	+2	+2	+2	**+2**	**+2**	**+2**	**+2**	+2	**+2**
+3	+3	+3	**+3**	+3	**+3**	+3	+3	+3	
	+4	+4	+4	+4	+4	+4	+4		
		+5	+5	+5	+5	+5			
			+6	+6	+6				
				+7					

- The variety of oxidation states results from the close similarity in energy of the 4s and the 3d electrons.
- The number of available oxidation states for the element **increases** from Sc → Mn.
  - All of the available 3d and 4s electrons may be used for bond formation.

Refer to the electronic configurations of the d-block elements, shown on page 63.

- The number of available oxidation states for the element **decreases** from Mn → Zn.
  - There is a decreasing number of unpaired d-electrons available for bond formation.
- Higher oxidation states involve covalency because of the high charge densities involved.
  - In practice, for oxidation states of +4 and above, the bonding electrons are involved in covalent bond formation.

Progress check

1. Write down the electronic configuration of the following ions:
   (a) Ni^{2+} (b) Cr^{3+} (c) Cu^+ (d) V^{3+}.

2. What is the oxidation state of the metal in each of the following?
   (a) MnO_2 (b) CrO_3 (c) MnO_4^- (d) VO_2^+

1 (a) [Ar]3d⁸ (b) [Ar]3d³ (c) [Ar]3d¹⁰ (d) [Ar]3d².
2 (a) +4 (b) +6 (c) +7 (d) +5.

The Periodic Table

3.3 Transition element complexes

After studying this section you should be able to:

- explain what is meant by the terms 'complex ion' and 'ligand'
- predict the formula and possible shape of a complex ion
- know that ligands can be unidentate, bidentate and multidentate
- understand that some complexes have stereoisomers
- know that transition metal ions can be identified by their colour
- know that electronic transitions are responsible for colour
- explain how colorimetry can be used to determine the concentration and formula of a complex ion

Ligands and complex ions

Transition metal ions are small and densely charged. They strongly attract electron-rich species called **ligands** forming **complex ions**.

> **KEY POINT**
> A ligand is a molecule or ion that bonds to a metal ion by:
> - forming a coordinate (dative covalent) bond
> - donating a lone pair of electrons into a vacant orbital.
> Common ligands include: $H_2O\!:\,$, $:\!Cl^-$, $:\!NH_3$, $:\!CN^-$

A ligand has a lone pair of electrons.

> **KEY POINT**
> A **complex ion** is a central metal ion surrounded by ligands.
> The **coordination number** is the total **number** of **coordinate bonds** from ligands to the central transition metal ion of a complex ion.

Key points from AS

- **Electron pair repulsion theory**
  Revise AS pages 45–46

For AS, you learnt that electron pair repulsion is responsible for the shape of a molecule or ion.

Shapes of complex ions

The size of a ligand helps to decide the shape of the complex ion.

Six coordinate complexes

Six water molecules are able to fit around a Cu^{2+} ion to form:

- the complex ion $[Cu(H_2O)_6]^{2+}$
- with a coordination number of **6**.

The **six** electron pairs surrounding the central Cu^{2+} ion in $[Cu(H_2O)_6]^{2+}$ repel one another as far apart as possible to form a complex ion with an **octahedral** shape.

The hexaaquacopper(II) complex ion, $[Cu(H_2O)_6]^{2+}$

$[Cu(H_2O)_6]^{2+}$
octahedral
6 coordinate

$$\left[\begin{array}{c} H_2O \\ H_2O\underset{H_2O}{\overset{|}{\diagdown}}\overset{90°}{\underset{|}{Cu}}\overset{\diagup}{\underset{OH_2}{OH_2}} \\ H_2O \end{array}\right]^{2+}$$

Four coordinate complexes

Chloride ions are larger than water molecules and it is only possible for **four** chloride ions to fit around the central copper(II) ion to form:

- the complex ion $[CuCl_4]^{2-}$
- with a coordination number of **4**.

The **four** electron pairs surrounding the central Cu^{2+} ion in $[CuCl_4]^{2-}$ repel one another as far apart as possible to form a complex ion with a **tetrahedral** shape.

The Periodic Table

The tetrachlorocuprate(II) complex ion $[CuCl_4]^{2-}$

Notice the overall charge on the complex ion $[CuCl_4]^{2-}$ is 2–.
Cu^{2+} and 4 Cl^- → 2– ions.

$[CuCl_4]^{2-}$ tetrahedral 4 coordinate

General rules for deciding the shape of a complex ion

Although there are exceptions, the following general rules are useful.

> Complex ions with small ligands such as H_2O and NH_3 are usually **6-coordinate** and **octahedral**.
>
> Complex ions with large Cl^- ligands are usually **4-coordinate** and **tetrahedral**.

KEY POINT

The table below compares complex ions formed from cobalt(II) and chromium(III) ions.

ligand	complex ions from Co^{2+}	complex ions from Cr^{3+}	shape
H_2O	$[Co(H_2O)_6]^{2+}$	$[Cr(H_2O)_6]^{3+}$	octahedral
NH_3	$[Co(NH_3)_6]^{2+}$	$[Cr(NH_3)_6]^{3+}$	octahedral
Cl^-	$[CoCl_4]^{2-}$	$[CrCl_4]^-$	tetrahedral

The $[Ag(NH_3)_2]^+$ complex ion present in Tollens' reagent (ammonical silver nitrate) is used as a test for the aldehyde functional group (see pages 94–95).

- Some complex ions contain more than one type of ligand. For example, copper(II) forms a complex ion with a mixture of water and ammonia ligands, $[Cu(NH_3)_4(H_2O)_2]^{2+}$.
- Ag^+ and Cu^+ forms linear complexes that are 2-coordinate: e.g. $[Ag(H_2O)_2]^+$, $[Ag(NH_3)_2]^+$, $[AgCl_2]^-$ and $[CuCl_2]^-$.

Ligands have teeth

Ligands such as H_2O, NH_3 and Cl^- are called **monodentate** ligands. They form only **one** coordinate bond to the central metal ion.

A molecule or ion with more than one oxygen or nitrogen atom may form more than one coordinate bond to the central metal ion.

Ligands that can form **two** coordinate bonds to the central metal ion are called **bidentate** ligands. Examples of bidentate ligands are ethane-1,2-diamine, $NH_2CH_2CH_2NH_2$ and the ethanedioate ion, $(COO^-)_2$.

Monodentate means **one** tooth.
Bidentate ligands have **two** teeth.
Multidentate ligands have **many** teeth.
Each 'tooth' is a coordinate bond.

Multidentate or polydentate ligands can form **many** coordinate bonds to the central metal ion. For example, the **hexadentate** ligand, edta^{4-}, is able to form **six** coordinate bonds to the central metal ion. The diagram of edta^{4-} shows four oxygen atoms and two nitrogen atoms able to form coordinate bonds.

edta^{4-}

Stereoisomerism in complexes

Stereoisomers have the same structural isomers but different arrangements in space. Transition metal complexes show two types of stereoisomerism:

- *cis-trans* isomerism
- optical isomerism.

The Periodic Table

Cis-trans isomerism

Cis-trans isomerism occurs in square planar complexes with two ligands of one kind and two of another kind, e.g. Ni(NH$_3$)$_2$Cl$_2$.

```
Cl      NH₃              H₃N      Cl
   \   /                     \   /
    Ni                         Ni
   /   \                     /   \
Cl      NH₃              Cl       NH₃

  cis-isomer              trans-isomer
groups on same side    groups on opposite side
```

Platin

The structures of *cis*-platin and *trans*-platin are shown below. Both structures have a square planar shape with a 90° bond angle.

```
Cl      NH₃              H₃N      Cl
   \   /                     \   /
    Pt                         Pt
   /   \                     /   \
Cl      NH₃              Cl       NH₃

  cis-platin               trans-platin
```

Cis-platin is used as a chemotherapy drug in the treatment of some cancers. *Cis*-platin forms coordinate bonds to a basic site in the DNA of cancer cells. This process prevents the cancerous DNA from replicating, effectively killing the cancerous cells. Careful treatment is required to avoid harmful side-effects of the drug.

Colour and complex ions

Complex ions have different colours

Many transition metal ions have different colours. These colours indicate the possible identity of a transition metal ion. The table below shows the colours of some **aqua** complex ions of d-block elements. Notice that different oxidation states of a d-block element can have different colours.

	Sc	Ti	V	Cr	Mn	Fe	Co	Ni	Cu	Zn
									+1 colourless	
					+2 pale pink	+2 pale green	+2 pink	+2 green	+2 blue	+2 colourless
	+3 colourless	+3 violet	+3 green	+3 *ruby / green		+3 red-brown				
			+5 yellow							
				+6 yellow or orange						
					+7 purple					

*Cr^{3+} surrounded by 6 water ligands looks violet, but some water is often replaced by other ligands so Cr^{3+} usually looks green.

The Periodic Table

Why are transition element ions coloured?

AQA M5

Splitting the 3d energy level

All five d-orbitals in the outer d sub-shell of an **isolated** transition element ion have the same energy. However, when ligands bond to the central ion, the d-orbitals move to two different energy levels. The gap between the energy levels depends upon:

- the ligand
- the coordination number
- the transition metal ion.

This diagram shows how water ligands split the energy levels of the five 3d orbitals in a copper(II) ion.

Absorption of light energy

An electron can be promoted from the lower 3d energy level by absorbing energy **exactly** equal to the energy gap ΔE. This energy is provided by radiation from the visible and ultraviolet regions of the spectrum.

Radiation with a frequency f has an energy $h\nu$, where h = Planck's constant, 6.63 × 10⁻³⁴ J s.

So, to promote an electron through as energy gap ΔE, a quantity of light energy is needed given by:

$$\Delta E = h\nu$$

Energy and frequency
The difference in energy between the d-orbitals, ΔE, and the frequency of the absorbed radiation, ν, are linked by the following relationship:
$\Delta E = h\nu$
h = Planck's constant, 6.63 × 10⁻³⁴ J s.

The diagram below shows what happens when a complex ion absorbs light energy.

Only those ions that have a partially filled d sub-shell are coloured. Cu^{2+} is $[Ar]3d^9 4s^2$

The blue colour of $[Cu(H_2O)_6]^{2+}$ results from:

- **absorption** of light energy in the red, yellow and green regions of the spectrum. This absorbed radiation provides the energy for an electron to be excited to a higher energy level
- **reflection** of blue light only, giving the copper(II) hexaaqua ion its characteristic blue colour. The blue region of the spectrum is **not** absorbed.

Copper has another ion, Cu^+, with the electronic configuration: $[Ar]3d^{10}$.

The Cu^+ ion is colourless because:

- the 3d sub-shell is full, $[Ar]3d^{10}$, preventing electron transfer.

Ag^+ complexes are also colourless because the 3d sub-shell is full.

> **KEY POINT**
> The colour of a transition metal complex ion results from the **transfer** of an **electron** between the orbitals of an **unfilled** d sub-shell.

The Periodic Table

Using colorimetry

AQA M5

A colorimeter measures the amount of light absorbed by a coloured solution. The concentration and formula of a complex ion can be determined from the intensity of light absorbed by the colorimeter.

Finding an unknown concentration of a coloured solution

- A suitable filter and wavelength is chosen for the readings.
- Solutions of a coloured complex ion of known concentration are prepared.
- The absorbances of the solutions of known concentration are measured.
- A calibration curve is plotted.
- The solution of the complex ion with an unknown concentration is then placed in the colorimeter and the absorbance measured.
- The unknown concentration can be determined from the calibration curve.

> Colorimetry can also be used as a method for monitoring reaction rates (see pages 13–18).

Progress check

1. Predict the formula of the following complex ions:
   (a) iron(III) with water ligands
   (b) vanadium(III) with chloride ligands
   (c) nickel(II) with ammonia ligands.
2. Explain the origin of colour in a complex ion that is yellow.

1 (a) $[Fe(H_2O)_6]^{3+}$ (b) $[VCl_4]^-$ (c) $[Ni(NH_3)_6]^{2+}$.
2 The 3d orbitals are split between two energy levels. A 3d electron from the lower 3d energy level absorbs visible light in the blue and red regions of the spectrum. The absorbed energy allows the d-electron to be promoted to a vacancy in the orbitals of the higher 3d energy level. With the red and blue regions of the spectrum absorbed, only yellow light is reflected, giving the complex ion its colour.

The Periodic Table

3.4 Ligand substitution of complex ions

After studying this section you should be able to:

- describe what is meant by ligand substitution
- understand that ligand exchange may produce changes in colour and coordination number
- understand stability of complex ions in terms of stability constants

Exchange between ligands

AQA M5

Ligand substitution usually produces a change in colour.

A **ligand substitution** reaction takes place when a ligand in a complex ion exchanges for another ligand. A change in ligand usually changes the energy gap between the 3d energy levels. With a different ΔE value, light with a different frequency is absorbed, producing a colour change.

Exchange between H_2O and NH_3 ligands

The **similar sizes** of water and ammonia molecules ensure that ligand exchange takes place with **no change in coordination number**.

For example, addition of an excess of concentrated aqueous ammonia to aqueous nickel(II) ions results in ligand substitution of ammonia ligands for water ligands. Both complex ions have an octahedral shape with a coordination number of six.

*Remember that the **size** of the ligand is a major factor in deciding the coordination number (see page 66):*

- *H_2O and NH_3 have similar sizes – same coordination number*
- *H_2O and Cl^- have different sizes – different coordination numbers.*

octahedral → same coordinate number → octahedral
6 coordinate → → 6 coordinate

$[Ni(H_2O)_6]^{2+}$ (green solution) + $6NH_3$ ⟶ $[Ni(NH_3)_6]^{2+}$ (blue solution) + $6H_2O$

Exchange between H_2O and Cl^- ligands

You should also be able to construct a balanced equation for any ligand exchange reaction.

The **different sizes** of water molecules and chloride ions ensure that ligand exchange takes place **with a change in coordination number**.

For example, addition of an excess of concentrated hydrochloric acid (as a source of Cl^- ions) to aqueous cobalt(II) ions results in ligand substitution of chloride ligands for water ligands. The complex ions have different shapes with different coordination numbers.

octahedral → change in coordinate number → tetrahedral
6 coordinate → → 4 coordinate

$[Co(H_2O)_6]^{2+}$ (blue solution) + $4Cl^-$ ⟶ $[CoCl_4]^{2-}$ (yellow solution) + $6H_2O$

The Periodic Table

Ligand substitutions of copper(II) and cobalt(II)

Ligand substitutions of copper(II) and cobalt(II) complexes are shown below.

> Concentrated hydrochloric acid is used as a source of Cl⁻ ligands.
> Concentrated aqueous ammonia is used as a source of NH₃ ligands.

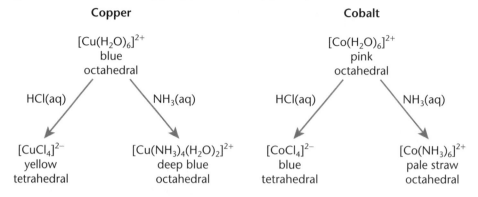

Incomplete substitution

Substitution may be incomplete.

- Aqueous ammonia only exchanges **four** from the six water ligands in $[Cu(H_2O)_6]^{2+}$.

$$[Cu(H_2O)_6]^{2+} + 4NH_3 \longrightarrow [Cu(NH_3)_4(H_2O)_2]^{2+} + 4H_2O$$
blue solution → deep blue solution

Haemoglobin

AQA M5

Haemoglobin is a complex ion of Fe^{2+}. A haem group consists of an Fe^{2+} ion, held in a ring, known as a porphyrin. The Fe^{2+} ion, which is the site of oxygen binding, bonds with four nitrogen atoms in the centre of the ring in a square planar geometry. The protein globin bonds below and O_2 is able to bond weakly above.

The porphyrin ring in haemoglobin

O_2 can be easily lost when required and is replaced by a weakly bonded water ligand. When haemoglobin reaches the lungs again, fresh O_2 is able to bond to Fe^{2+}, displacing the water.

Carbon monoxide, CO, is a serious poison for the blood as the polar molecule binds much more strongly to the Fe^{2+} ion than O_2 and cannot be displaced. This prevents oxygen from being carried by the blood.

The Periodic Table

Stability of complex ions

AQA M5

The stability of a complex ion depends upon the ligands. For example, a complex ion of copper(II) is more stable with ligands of ammonia than with water ligands.

Bidentate or multidentate ligands (such as edta) lead to a particularly stable complex. This is due to an increase in entropy:

e.g. $\underbrace{[Cu(H_2O)_6]^{2+} + edta^{4-}}_{2 \text{ mol}} \xrightarrow{\text{INCREASE IN ENTROPY}} \underbrace{[Cu(edta)]^{2-} + 6H_2O}_{7 \text{ mol}}$

Progress check

1. Write equations for the following ligand substitutions:
   (a) $[Ni(H_2O)_6]^{2+}$ with six ammonia ligands
   (b) $[Fe(H_2O)_6]^{3+}$ with four chloride ligands
   (c) $[Cr(H_2O)_6]^{3+}$ with six cyanide ligands, CN^-.

1 (a) $[Ni(H_2O)_6]^{2+} + 6NH_3 \longrightarrow [Ni(NH_3)_6]^{2+} + 6H_2O$
   (b) $[Fe(H_2O)_6]^{3+} + 4Cl^- \longrightarrow [FeCl_4]^- + 6H_2O$
   (c) $[Cr(H_2O)_6]^{3+} + 6CN^- \longrightarrow [Cr(CN)_6]^{3-} + 6H_2O$.

The Periodic Table

3.5 Redox reactions of transition metal ions

After studying this section you should be able to:

- describe redox behaviour in transition elements
- construct redox equations, using relevant half-equations
- perform calculations involving simple redox titrations

LEARNING SUMMARY

Redox reactions

AQA M5

Transition elements have variable oxidation states with characteristic colours (see page 68). Many **redox** reactions take place in which transition metals ions change their oxidation state by **gaining** or **losing** electrons.

The iron(II) – manganate(VII) reaction

Key points from AS
- Redox reactions
  Revise AS pages 33–35

Iron(II) ions are oxidised by acidified manganate(VII) ions.

- Acidified manganate(VII) ions are reduced to manganate(II) ions.

reduction: $MnO_4^-(aq) + 8H^+(aq) + 5e^- \longrightarrow Mn^{2+}(aq) + 4H_2O(l)$
              +7   +2
              purple $\longrightarrow$ very pale pink

The reaction is easy to see.

The deep purple MnO_4^- ions are reduced to the very pale pink Mn^{2+} ions, which are virtually colourless in solution.

- Iron(II) ions are oxidised to iron(III) ions.

oxidation: $Fe^{2+}(aq) \longrightarrow Fe^{3+}(aq) + e^-$
              +2   +3

To give the overall equation:

- the electrons are balanced

$$MnO_4^-(aq) + 8H^+(aq) + 5e^- \longrightarrow Mn^{2+}(aq) + 4H_2O(l)$$
$$5Fe^{2+}(aq) \longrightarrow 5Fe^{3+}(aq) + 5e^-$$

- the half-equations are added

$$5Fe^{2+}(aq) + MnO_4^-(aq) + 8H^+(aq) \longrightarrow 5Fe^{3+}(aq) + Mn^{2+}(aq) + 4H_2O(l)$$

Redox titrations

AQA M5

Redox titrations can be used in analysis.

Essential requirements are:

- an oxidising agent
- a reducing agent
- an easily-seen colour change (with or without an indicator).

*It is important to use an acid that does **not** react with either the oxidising or reducing agent – dilute sulphuric acid is usually used.*

For example, hydrochloric acid cannot be used because HCl is oxidised to Cl_2 by MnO_4^-.

Redox titrations using the acidified manganate(VII)

- $KMnO_4$ is reacted with a reducing agent such as Fe^{2+} under acidic conditions (see above).
- Purple $KMnO_4$ is added from the burette to a measured amount of the acidified reducing agent.
- At the end-point, the colour changes from colourless to pale pink, showing that all the reducing agent has exactly reacted. The pale-pink colour indicates the first trace of an excess of purple manganate(VII) ions.
- This titration is **self-indicating** – no indicator is required, the colour change from the reduction of MnO_4^- ions being sufficient to indicate that the reaction is complete.

The most common redox titration encountered at A Level is that between manganate(VII) and acidified iron(II) ions.

The Periodic Table

> **Key points from AS**
> - Calculations in acid–base titrations
>   *Revise AS page 32*

The estimation of iron in iron tablets

Five iron tablets with a combined mass of 0.900 g were dissolved in acid and made up to 100 cm³ of solution. In a titration, 10.0 cm³ of this solution reacted exactly with 10.4 cm³ of 0.0100 mol dm⁻³ potassium manganate(VII). What is the percentage by mass of iron in the tablets?

As with all titrations, we must consider **five** pieces of information:

- the balanced equation
  $$5Fe^{2+}(aq) + MnO_4^-(aq) + 8H^+(aq) \rightarrow 5Fe^{3+}(aq) + Mn^{2+}(aq) + 4H_2O(l)$$
- the concentration c_1 and reacting volume V_1 of $KMnO_4(aq)$
- the concentration c_2 and reacting volume V_2 of $Fe^{2+}(aq)$.

> Calculations for redox titrations follow the same principles as those used for acid–base titrations, studied during AS Chemistry.

From the titration results, the amount of $KMnO_4$ can be calculated:

$$\text{amount of } KMnO_4 = c \times \frac{V}{1000} = 0.0100 \times \frac{10.4}{1000} = 1.04 \times 10^{-4} \text{ mol}$$

> The concentration and volume of $KMnO_4$ are known so the amount of MnO_4^- ions can be found.

From the equation, the amount of Fe^{2+} can be determined:

$$5Fe^{2+}(aq) + MnO_4^-(aq) + 8H^+(aq) \rightarrow 5Fe^{3+}(aq) + Mn^{2+}(aq) + 4H_2O(l)$$

     5 mol   1 mol   (balancing numbers)

∴ $5 \times 1.04 \times 10^{-4}$ mol Fe^{2+} produced from 1.04×10^{-4} mol MnO_4^-

∴ amount of Fe^{2+} that reacted $= 5.20 \times 10^{-4}$ mol

> Using the equation determine the number of moles of the second reagent.

Find the amount of Fe^{2+} in the solution prepared from the tablets:

10.0 cm³ of $Fe^{2+}(aq)$ contains 5.20×10^{-4} mol Fe^{2+} (aq)
The 100 cm³ solution of iron tablets contains $10 \times (5.20 \times 10^{-4})$
$$= 5.20 \times 10^{-3} \text{ mol } Fe^{2+}$$

> If there is sampling of the original solution, you will need to scale.

Find the percentage of Fe^{2+} in the tablets (A_r: Fe, 55.8)

5.20×10^{-3} mol Fe^{2+} has a mass of $5.20 \times 10^{-3} \times 55.8 = 0.290$ g

$$\therefore \% \text{ of } Fe^{2+} \text{ in tablets} = \frac{\text{mass of } Fe^{2+}}{\text{mass of tablets}} \times 100 = \frac{0.290}{0.900} \times 100 = 32.2\%$$

Further redox titrations

> For A2, you are expected to tackle unstructured titration calculations. The principles are the same, irrespective of the type of titration.

Providing there is a visible colour change, many other redox reactions can be used for redox titrations.

For example, acidified dichromate(VI) ions is an oxidising agent which can also oxidise Fe^{2+} ions to Fe^{3+}.

reduction: $Cr_2O_7^{2-}(aq) + 14H^+(aq) + 6e^- \longrightarrow 2Cr^{3+}(aq) + 7H_2O(l)$
oxidation: $Fe^{2+}(aq) \longrightarrow Fe^{3+}(aq) + e^-$
overall: $6Fe^{2+}(aq) + Cr_2O_7^{2-}(aq) + 14H^+(aq) \rightarrow 6Fe^{3+}(aq) + 2Cr^{3+}(aq) + 7H_2O(l)$

Further examples of redox reactions

Vanadium

Vanadium(V) can be reduced to vanadium(II) using zinc in acid solution as the reducing agent. The formation of different oxidation states of vanadium can be followed by colour changes.

oxidation state	+2	+3	+4	+5
species	V^{2+} (violet)	V^{3+} (green)	VO^{2+} (blue)	VO_2^+ (yellow)

                                  ⟵ Zn/HCl

The Periodic Table

Chromium

The oxidation states of chromium can be interchanged using:

- hydrogen peroxide in alkaline solution as the oxidising agent
- zinc in acid solution as the reducing agent.

> For moving **up** oxidation states H_2O_2/OH^- is a good general oxidising agent.
>
> For moving **down** oxidation states Zn/HCl is a good general reducing agent.

The dichromate(VI) – chromate(VI) conversion

The table on the previous page shows that the dichromate(VI) ion $Cr_2O_7^{2-}$ and the chromate(VI) ion CrO_4^{2-} have the **same** oxidation state. These ions can be interconverted at different pH values.

$$2H^+(aq) + 2CrO_4^{2-}(aq) \underset{\text{alkali}}{\overset{\text{acid}}{\rightleftharpoons}} Cr_2O_7^{2-}(aq) + H_2O(l)$$

yellow → orange

> This conversion can be explained by applying le Chatelier's principle.
>
> Addition of acid increases $[H^+(aq)]$ – the equilibrium moves to the right, counteracting this change.
>
> Added alkali reacts with $H^+(aq)$ ions present in the equilibrium system. $[H^+(aq)]$ decreases – the equilibrium moves to the left, counteracting this change.

Cobalt

Cobalt(II) can be oxidised to Co(III) using:

- hydrogen peroxide in alkaline solution or
- air in the presence of ammonia as an oxidising agent.

oxidation state	+2		+3
species	$[Co(NH_3)_6]^{2+}$ (light straw)	H_2O_2/OH^- or NH_3/air $\longrightarrow$	$[Co(NH_3)_6]^{3+}$ (brown)

Progress check

1. Write a full equation from the following pairs of half-equations. For each pair, identify the changes in oxidation number, what has been oxidised and what has been reduced.
   (a) $Zn \longrightarrow Zn^{2+} + 2e^-$
   $VO_2^+ + 4H^+ + 3e^- \longrightarrow V^{2+} + 2H_2O$
   (b) $Cr^{3+} + 8OH^- \longrightarrow CrO_4^{2-} + 4H_2O + 3e^-$
   $H_2O_2 + 2e^- \longrightarrow 2OH^-$

2. In a redox titration, 25.0 cm³ of an acidified solution containing $Fe^{2+}(aq)$ ions reacted exactly with 21.8 cm³ of 0.0200 mol dm⁻³ potassium manganate(VII). Calculate the concentration of $Fe^{2+}(aq)$ ions.

1 (a) $2VO_2^+ + 3Zn + 8H^+ \longrightarrow 2V^{2+} + 3Zn^{2+} + 4H_2O$
V: +5 → +2; Zn: 0 → +2; VO_2^+ reduced; Zn oxidised.
(b) $2Cr^{3+} + 3H_2O_2 + 10OH^- \longrightarrow 2CrO_4^{2-} + 8H_2O$
Cr: +3 → +6; O: −1 → −2; Cr^{3+} oxidised; H_2O_2 reduced.
2 0.0872 mol dm⁻³

The Periodic Table

3.6 Catalysis

After studying this section you should be able to:

- understand how a transition element acts as a catalyst
- explain how a catalyst acts in heterogeneous catalysis
- explain how a catalyst acts in homogeneous catalysis

Key points from AS

- How do catalysts work?
  Revise AS page 87

The two different classes of catalyst, homogeneous and heterogeneous, were discussed in detail during AS Chemistry. Both types are used industrially but heterogeneous catalysis is much more common. The AS coverage is summarised below and expanded to include extra examples of transition element catalysis.

Transition elements as catalysts

AQA   M5

An important use of transition metals and their compounds is as catalysts for many industrial processes. Nickel and platinum are extensively used in the petroleum and polymer industries.

> **KEY POINT**
>
> Transition metal ions are able to act as catalysts by changing their oxidation states. This is made possible by using the partially full d-orbitals for gaining or losing electrons.

Heterogeneous catalysis

AQA   M5

A **heterogeneous** catalyst has a **different** phase from the reactants.

- Many examples of heterogeneous catalysts involve reactions between **gases**, catalysed by a **solid** catalyst which is often a transition metal or one of its compounds.

The process involves:

- **diffusion** of gas molecules onto the surface of the iron catalyst
- **adsorption** of the gases to the surface of the catalyst
- **weakening of bonds**, allowing a chemical reaction to take place
- **diffusion** of the product molecules from the surface of the catalyst, allowing more gas molecules to diffuse onto its surface.

Transition metals use d and s-electrons to form weak bonds with reactant molecules at the catalyst surface.

Transition metal ions as heterogeneous catalysts

The catalytic converter

- Rh/Pt/Pd catalysts are used in catalytic converters for the removal of polluting gases, such as CO and NO, produced in a car engine.
- The **solid** Rh/Pt/Pd catalyst is held on a ceramic honeycomb support, giving a large surface area for the catalyst which can be spread extremely thinly on the support. The high surface area of the catalyst increases the chances of a reaction and less of the expensive catalysts are needed, keeping down costs.

Iron

Iron-containing catalysts are used in many industrial processes, the most important being to form ammonia from nitrogen and hydrogen in the Haber process.

$$N_2(g) + 3H_2(g) \rightleftharpoons 2NH_3(g)$$

The Periodic Table

Chromium(III) oxide

Chromium(III) oxide, Cr_2O_3, is used to catalyse the manufacture of methanol from carbon monoxide and hydrogen.

$$CO(g) + 2H_2(g) \rightleftharpoons CH_3OH(g)$$

Vanadium

A vanadium(V) oxide, V_2O_5, catalyst is used in the contact process for sulfur trioxide production from which sulfuric acid is manufactured.

This catalysis proceeds via an **intermediate state**.

- The vanadium(V) oxide catalyst first oxidises sulfur dioxide to sulfur trioxide. The vanadium is **reduced** from the +5 to +4 oxidation state, forming vanadium(IV) oxide $V_2O_4(s)$ as the **intermediate state**.

$$SO_2(g) + V_2O_5(s) \longrightarrow SO_3(g) + V_2O_4(s) \quad \text{vanadium reduced}$$
+4 $\longrightarrow$ +6
+5 $\longrightarrow$ +4
+5 $\longrightarrow$ +4

> Notice that the oxidation number of each atom has been shown. Two V atoms have been reduced from +5 to +4.

- The intermediate V_2O_4 is then oxidised back to the +5 oxidation state by oxygen, forming vanadium(V) oxide.

$$½O_2(g) + V_2O_4(s) \longrightarrow V_2O_5(s) \quad \text{vanadium oxidised}$$
0 $\longrightarrow$ −2
+4 $\longrightarrow$ +5
+4 $\longrightarrow$ +5

This is a chain reaction because the vanadium(V) oxide is then able to oxidise further sulfur dioxide.

Note:

- Although V_2O_5 has been involved in the mechanism, the overall reaction does not include V_2O_5. V_2O_5 is unchanged at the end of the reaction.
- The equation for the overall reaction taking place is:
  $$SO_2(g) + ½O_2(g) \rightleftharpoons SO_3(g) \quad \text{The contact process}$$
- The vanadium is able to catalyse the reaction by interchanging its +5 and +4 oxidation states.

Poisoning of catalysts

A key requirement of a heterogeneous catalyst is to bind reactant molecules to the surface of the catalyst by adsorption.

The strength of this adsorption is critical.

- Lead bonds very strongly to the catalyst. The adsorption is so strong that molecules are unable to leave the surface and the catalyst is poisoned and rendered useless. Before catalytic converters could be introduced, unleaded petrol was developed. Otherwise, the lead in the fuel would poison the catalyst.

> Catalysts can become coated by impurities and can become 'poisoned' – efficiency is reduced as active sites are blocked.

Nickel and platinum are commonly used as heterogeneous catalysts.

- The bonding of reactant molecules is sufficient to allow a reaction to take place but weak enough to allow molecules to escape following reaction.

Homogeneous catalysis

AQA M5

A **homogeneous catalyst** has the **same** phase as the reactants.

> **Key points from AS**
> - How catalysts work
>   Revise AS page 87

- During homogeneous catalysis, the catalyst initially reacts with the reactants forming an **intermediate state** with **lower activation energy**.

The Periodic Table

- The catalyst is then regenerated as the reaction completes, allowing the catalyst to react with further reactants.
- Overall the catalyst is **not** used up, but it is actively involved in the process by a chain reaction.

Transition metal ions as homogeneous catalysts

The reaction between I^- and $S_2O_8^{2-}$, catalysed by $Fe^{2+}(aq)$ ions

The reaction between I^- and $S_2O_8^{2-}$ is very slow:

- it can be catalysed by Fe^{2+} ions
- the Fe^{2+} ions are recycled in a chain reaction.

chain reaction: $\quad S_2O_8^{2-}(aq) + 2Fe^{2+}(aq) \longrightarrow 2SO_4^{2-}(aq) + 2Fe^{3+}(aq)$

$\quad\quad\quad\quad\quad\quad\quad\quad 2Fe^{3+}(aq) + 2I^-(aq) \longrightarrow I_2(aq) + 2Fe^{2+}(aq)$

overall: $\quad\quad\quad\quad S_2O_8^{2-}(aq) + 2I^-(aq) \longrightarrow I_2(aq) + 2SO_4^{2-}(aq)$

> Transition metal ions often change their oxidation states during their catalytic action.
> Here: $Fe^{2+} \longrightarrow Fe^{3+}$

The overall reaction between I^- and $S_2O_8^{2-}$ involves two negative ions. These negative ions will repel one another, making it difficult for a reaction to take place. Notice that, in each stage of the mechanism, the catalyst provides positive ions which attract, and react with, each negative ion in turn.

Autocatalysis by $Mn^{2+}(aq)$ ions

> Notice that, as in the first example, this reaction is between two negative ions.
> The mechanism of this reaction proceeds in a similar manner – the catalyst provides positive ions which react with each negative ion.

In warm acidic conditions, manganate(VII) ions, MnO_4^-, oxidise ethanedioate ions, $C_2O_4^{2-}$, to carbon dioxide.

$$2MnO_4^-(aq) + 5C_2O_4^{2-}(aq) + 8H^+(aq) \rightarrow 2Mn^{2+}(aq) + 10CO_2(g) + H_2O(l)$$

The purple manganate(VII) colour disappears very slowly at first, indicating a slow reaction. As soon as $Mn^{2+}(aq)$ ions start to form, the colour disappears immediately because the $Mn^{2+}(aq)$ ions catalyse the reaction. This is called **autocatalysis**.

Progress check

1. (a) State what is meant by homogeneous and heterogeneous catalysis.
   (b) State an example of a homogeneous and heterogeneous catalyst.
2. How does a transition metal act as a catalyst?

1 (a) Homogeneous: catalyst and reactants have the same phase.
Heterogeneous: catalyst and reactants have different phases.
(b) Homogeneous: Fe^{2+} catalysing the reaction between I^- and $S_2O_8^{2-}$.
Heterogeneous: iron catalysing the reaction between N_2 and H_2 for ammonia production (Haber process).
2 Transition metal changes oxidation states by gaining or losing electrons from partially filled d-orbitals.

The Periodic Table

3.7 Reactions of metal aqua-ions

After studying this section you should be able to:

- describe the precipitation reactions of metal aqua-ions with bases
- describe the acidity of transition metal aqua-ions
- describe the acid–base behaviour of metal hydroxides

Simple precipitation reactions of metal aqua-ions

AQA M5

*Notice that NaOH(aq) and NH$_3$(aq) **both** act as bases.*

A precipitation reaction takes place between **aqueous alkali** and an aqueous solution of a **metal(II)** or **metal(III) cation**.

This results in the formation of a **precipitate** of the **metal hydroxide**, often with a characteristic colour.

Suitable aqueous alkalis include aqueous sodium hydroxide, NaOH(aq), and aqueous ammonia, NH$_3$(aq).

The characteristic colour of the precipitate can help to identify the metal ion. The colours of some hydroxide precipitates are shown below.

hydroxide	Fe(OH)$_2$(s)	Fe(OH)$_3$(s)	Co(OH)$_2$(s)	Cu(OH)$_2$(s)	Cr(OH)$_3$(s)
colour	green	brown	blue–green	blue	green

Reaction of complex metal aqua-ions with aqueous alkali

See also:
'Reactions of metal aqua-ions with aqueous bases' page 82.

The precipitate has no charge.

Equations can be written using complex aqua-ions for the reactions above, each producing a **precipitate** of the **hydrated hydroxide**. This precipitation is an **acid–base** reaction (see page 81).

$[Cu(H_2O)_6]^{2+}(aq) + 2OH^-(aq) \longrightarrow Cu(OH)_2(H_2O)_4(s) + 2H_2O(l)$
  blue precipitate } reactions with NaOH(aq)

$[Cr(H_2O)_6]^{3+}(aq) + 3OH^-(aq) \longrightarrow Cr(OH)_3(H_2O)_3(s) + 3H_2O(l)$
  green precipitate

$[Cu(H_2O)_6]^{2+}(aq) + 2NH_3(aq) \longrightarrow Cu(OH)_2(H_2O)_4(s) + 2NH_4^+(aq)$ } reaction with NH$_3$(aq)
  blue precipitate

Excess aqueous sodium hydroxide

Metal(III) hydroxides **dissolve** in an **excess** of dilute aqueous sodium hydroxide forming charged **anions**.

$Al(OH)_3(H_2O)_3(s) + OH^-(aq) \rightleftharpoons [Al(OH)_4]^-(aq) + 3H_2O(l)$
  aluminate ions

Species in solution are charged.

$Cr(OH)_3(H_2O)_3(s) + 3OH^-(aq) \rightleftharpoons [Cr(OH)_6]^{3-}(aq) + 3H_2O(l)$
  chromate(III) ions

Excess aqueous ammonia

Excess ammonia usually results in **ligand exchange** forming a soluble ammine complex:

$Cu(OH)_2(H_2O)_4(s) + 4NH_3(aq) \longrightarrow [Cu(NH_3)_4(H_2O)_2]^{2+}(aq) + 2H_2O(l) + 2OH^-(aq)$

$Cr(H_2O)_3(OH)_3(s) + 6NH_3(aq) \longrightarrow [Cr(NH_3)_6]^{3+}(aq) + 3H_2O(l) + 3OH^-(aq)$
  purple solution

If aqueous ammonia is slowly added to an aqueous solution of a **metal(II)** or **metal(III) cation**:

- a precipitate of the metal hydroxide first forms: acid–base reaction
- the precipitate then dissolves in excess ammonia: ligand substitution.

The Periodic Table

Precipitation as acid–base equilibria

AQA   M5

Fission of O–H bond

Metal(II) aqua-ions are weaker acids than metal(III) aqua-ions.

Examples of metal aqua-ions with 3+ cations include $[Al(H_2O)_6]^{3+}$, $[V(H_2O)_6]^{3+}$, $[Cr(H_2O)_6]^{3+}$ and $[Fe(H_2O)_6]^{3+}$.

Acidity of metal aqua-ions

Metal(II) and metal(III) aqua-ions behave as weak acids. This is achieved by:
- fission of the O–H bond in one of the water ligands of the complex ion
- donation of a proton to a water molecule.

Metal(II) cations

- With a metal(II) aqua-ion, the equilibrium lies well to the left-hand side: only a very weak acid is formed.

$$[Cu(H_2O)_6]^{2+}(aq) + H_2O(l) \rightleftharpoons [Cu(H_2O)_5(OH)]^+(aq) + H_3O^+(l)$$
very weakly acidic

Metal(III) cations

- Metal(III) aqua-ions are slightly more acidic than metal(II) aqua-ions: the equilibrium position has moved slightly to the right.

$$[Fe(H_2O)_6]^{3+}(aq) + H_2O(l) \rightleftharpoons [Fe(H_2O)_5(OH)]^{2+}(aq) + H_3O^+(aq)$$
weakly acidic

> **KEY POINT** The aqua-ions of metal(III) cations have a greater charge/size ratio and greater acidity than aqua-ions of metal(II) cations.

Precipitation

The precipitation reaction of a transition metal with hydroxide ions can be expressed as a series of acid–base equilibria.

The equilibrium set up in water by the hexaaqua ion of iron(III) is:

$$[Fe(H_2O)_6]^{3+}(aq) + H_2O(l) \rightleftharpoons [Fe(H_2O)_5(OH)]^{2+}(aq) + H_3O^+(aq)$$

OH⁻ reacts with H_3O^+:
$OH^- + H_3O^+ \longrightarrow 2H_2O$

With the weak alkali $Na_2CO_3(aq)$, metal(II) aqua-ions precipitate metal **carbonates**.

The more acidic metal(III) aqua-ions precipitate metal **hydroxides**.

Metal(III) carbonates do not exist.

On addition of aqueous hydroxide ions:
- H_3O^+ ions are removed from the equilibrium, which shifts to the right.

The $[Fe(H_2O)_5(OH)]^{2+}$ ion now sets up a second equilibrium:

$$[Fe(H_2O)_5(OH)]^{2+}(aq) + H_2O(l) \rightleftharpoons [Fe(H_2O)_4(OH)_2]^+(aq) + H_3O^+(aq)$$

- Again, the hydroxide ion removes H_3O^+ ions from the aqueous equilibrium, which shifts to the right.
- This process continues. Eventually **all charge** will be **removed** from the iron(III) complex. Then the uncharged **hydrated metal hydroxide** precipitates.

$$[Fe(H_2O)_4(OH)_2]^+(aq) + H_2O(l) \rightleftharpoons [Fe(H_2O)_3(OH)_3](s) + H_3O^+(aq)$$

Acid–base properties of metal hydroxides

AQA   M5

See also:
'Reaction of complex metal aqua-ions with aqueous alkali' page 80.

Many insoluble metal hydroxides are **amphoteric** – they can act as both an acid and a base.

Metal hydroxides as acids

Some insoluble metal hydroxides can act as acids, donating protons to strong alkalis and forming soluble anions.

- Metal(III) hydroxides react with dilute aqueous sodium hydroxide.

$$Al(OH)_3(H_2O)_3(s) + OH^-(aq) \rightleftharpoons [Al(OH)_4(H_2O)_2]^-(aq) + H_2O(l)$$
aluminate ions

$$Cr(OH)_3(H_2O)_3(s) + 3OH^-(aq) \rightleftharpoons [Cr(OH)_6]^{3-}(aq) + 3H_2O(l)$$
chromate(III) ions

The Periodic Table

In general, metal(III) hydroxides form anions with the general formula $[M(OH)_6]^{3-}$.

- Metal(II) hydroxides are weaker acids and do not usually react with aqueous sodium hydroxide.

Metal hydroxides as bases

All metal hydroxides can act as bases, removing protons from strong acids.

$$Cu(OH)_2(H_2O)_4(s) + 2H^+(aq) \rightleftharpoons [Cu(H_2O)_6]^{2+}(aq)$$
blue precipitate

This reaction reverses the direction of the acid–base equilibria that produce hydrated hydroxide precipitates (see page 81).

$$Cr(OH)_3(H_2O)_3(s) + 3H^+(aq) \rightleftharpoons [Cr(H_2O)_6]^{3+}(aq)$$
green precipitate

Reactions of metal aqua-ions with aqueous bases

AQA M5

Reactions with OH⁻(aq) and NH₃(aq)

You are expected to know reactions of metal aqua ions with aqueous $OH^-(aq)$ ions (probably as NaOH(aq)) and with aqueous ammonia, $NH_3(aq)$.

The tables below summarise what you are expected to know.

In exams, a great emphasis is placed on these reactions. You will need to learn the colours of all solutions and precipitates in your course.

	AQA
Fe^{2+}	✓
Co^{2+}	✓
Cu^{2+}	✓
Cr^{3+}	✓
Fe^{3+}	✓
Al^{3+}	✓
OH^-	✓
OH^- excess	✓
NH_3	✓
NH_3 excess	✓

aqua ion	OH⁻(aq) or NH₃(aq) as a base		NH₃(aq) as a ligand
	NaOH(aq) or dilute NH₃(aq) (small amount)	excess NaOH(aq)	excess/conc. NH₃(aq)
$[Fe(H_2O)_6]^{2+}$ green	$Fe(OH)_2(s)$ / $Fe(OH)_2(H_2O)_4(s)$ green	no change	no change
$[Co(H_2O)_6]^{2+}$ pink	$Co(OH)_2(s)$ / $Co(OH)_2(H_2O)_4(s)$ blue–green	no change	$[Co(NH_3)_6]^{2+}$ pale straw
$[Cu(H_2O)_6]^{2+}$ blue	$Cu(OH)_2(s)$ / $Cu(OH)_2(H_2O)_4(s)$ pale blue	no change	$[Cu(NH_3)_4(H_2O)_2]^{2+}$ deep blue
$[Cr(H_2O)_6]^{3+}$ ruby	$Cr(OH)_3(s)$ / $Cr(OH)_3(H_2O)_3(s)$ green	$[Cr(OH)_6]^{3-}(aq)$ green	$[Cr(NH_3)_6]^{3+}(aq)$ purple
$[Fe(H_2O)_6]^{3+}$ red-brown	$Fe(OH)_3(s)$ / $Fe(OH)_3(H_2O)_3(s)$ brown	no change	no change
$[Al(H_2O)_6]^{3+}$ colourless	$Al(OH)_3(s)$ / $Al(OH)_3(H_2O)_3(s)$ white	$[Al(OH)_4(H_2O)_2]^-(aq)$ colourless	no change

The Periodic Table

- $Fe(OH)_2(s)$ is oxidised in air forming a brown precipitate of $Fe(OH)_3(s)$.
- $[Co(NH_3)_6]^{2+}$ darkens in air as it is oxidised to $[Co(NH_3)_6]^{3+}$ (see page 76).

Reactions of metal aqua-ions with aqueous carbonate ions

Metal(II) ions

Metal(II) ions form a precipitate of the metal carbonate. The chemistry is complex and the equations here are acceptable simplifications.

$$Fe^{2+}(aq) + CO_3^{2-}(aq) \longrightarrow FeCO_3(s)$$
green

$$Co^{2+}(aq) + CO_3^{2-}(aq) \longrightarrow CoCO_3(s)$$
pink

$$Cu^{2+}(aq) + CO_3^{2-}(aq) \longrightarrow CuCO_3(s)$$
green

Metal(III) ions

Metal(III) aqua-ions behave as weak acids (see page 81).

A hydroxide precipitate is formed together with CO_2 gas.

$$[Cr(H_2O)_6]^{3+} + 1\tfrac{1}{2}CO_3^{2-}(aq) \longrightarrow Cr(OH)_3(H_2O)_3(s) + 1\tfrac{1}{2}CO_2(g) + 1\tfrac{1}{2}H_2O(l)$$
green

$$[Fe(H_2O)_6]^{3+} + 1\tfrac{1}{2}CO_3^{2-}(aq) \longrightarrow Fe(OH)_3(H_2O)_3(s) + 1\tfrac{1}{2}CO_2(g) + 1\tfrac{1}{2}H_2O(l)$$
brown

$$[Al(H_2O)_6]^{3+} + 1\tfrac{1}{2}CO_3^{2-}(aq) \longrightarrow Al(OH)_3(H_2O)_3(s) + 1\tfrac{1}{2}CO_2(g) + 1\tfrac{1}{2}H_2O(l)$$
white

- The acidity of the metal(III) ion means that this reaction is essentially the same as that between an acid such as hydrochloric acid and a carbonate.
- Consequently, metal(III) carbonates do not exist.

Progress check

1. Write equations for the following reactions of metal aqua-ions with aqueous alkali. What is the colour of each precipitate?

   (a) $[Co(H_2O)_6]^{2+}(aq)$ (b) $[Fe(H_2O)_6]^{2+}(aq)$ (c) $[Ni(H_2O)_6]^{2+}(aq)$
   (d) $[Fe(H_2O)_6]^{3+}(aq)$.

2. Write equations to show what happens when excess $NH_3(aq)$ is added to the following hydrated metal hydroxides.

   (a) $Co(OH)_2(H_2O)_4$ (b) $Cu(OH)_2(H_2O)_4$ (c) $Cr(OH)_3(H_2O)_3$.

---

1. (a) $[Co(H_2O)_6]^{2+}(aq) + 2OH^-(aq) \longrightarrow Co(OH)_2(H_2O)_4(s) + 2H_2O(l)$ (blue green)
   (b) $[Fe(H_2O)_6]^{2+}(aq) + 2OH^-(aq) \longrightarrow Fe(OH)_2(H_2O)_4(s) + 2H_2O(l)$ (green)
   (c) $[Ni(H_2O)_6]^{2+}(aq) + 2OH^-(aq) \longrightarrow Ni(OH)_2(H_2O)_4(s) + 2H_2O(l)$ (pale green)
   (d) $[Fe(H_2O)_6]^{3+}(aq) + 3OH^-(aq) \longrightarrow Fe(OH)_3(H_2O)_3(s) + 3H_2O(l)$ (brown)

2. (a) $Co(OH)_2(H_2O)_4(s) + 6NH_3(aq) \longrightarrow [Co(NH_3)_6]^{2+}(aq) + 4H_2O(l) + 2OH^-(aq)$ (pale straw)
   (b) $Cu(OH)_2(H_2O)_4(s) + 4NH_3(aq) \longrightarrow [Cu(NH_3)_4(H_2O)_2]^{2+}(aq) + 2H_2O(l) + 2OH^-(aq)$ (deep-blue)
   (c) $Cr(OH)_3(H_2O)_3(s) + 6NH_3(aq) \longrightarrow [Cr(NH_3)_6]^{3+}(aq) + 3H_2O(l) + 3OH^-(aq)$ (purple)

The Periodic Table

Sample question and model answer

This question looks at different transition elements.

(a) Write the electron configuration of a vanadium atom and a cobalt(III) ion.

V: $1s^2 2s^2 2p^6 3s^2 3p^6 3d^3 4s^2$ ✓

Co^{3+}: $1s^2 2s^2 2p^6 3s^2 3p^6 3d^6$ ✓ [2]

Remember:
4s fill before 3d
4s empty before 3d

(b) When concentrated hydrochloric acid is added to an aqueous solution containing copper(II) ions, the colour changes. Write the formulae of the two copper complexes involved, their colours, their shapes and the co-ordination numbers of the copper. Write an equation for the reaction and explain why the addition of dilute hydrochloric acid does not produce the same colour change.

Aqueous copper(II) contains the complex $[Cu(H_2O)_6]^{2+}$ ✓ which is pale blue. ✓
It has an octahedral shape ✓ and a coordination number of 6. ✓
After addition of concentrated hydrochloric acid, the complex ion $[CuCl_4]^{2-}$ ✓ is formed, which is yellow. ✓ It has a tetrahedral shape ✓ and a coordination number of 4. ✓

Equation: $[Cu(H_2O)_6]^{2+} + 4Cl^- \longrightarrow CuCl_4^{2-} + 6H_2O$ ✓✓

The concentration of chloride ligands in **dilute** hydrochloric acid is much smaller than the concentration of water ligands. ✓ [11]

This is standard bookwork but it also reinforces important principles:
H_2O forms octahedral complexes with a coordination number of 6.
Cl^- forms tetrahedral complexes with a coordination number of 4.

(c) A 0.626 g sample of hydrated iron(II) sulfate was dissolved in water and the solution made up to 250 cm³. A 25.0 cm³ sample of this solution was acidified with dilute sulfuric acid and titrated with potassium dichromate(VI). 25.85 cm³ of 0.00145 mol dm⁻³ potassium dichromate(VI) were required for exact reaction.

The half-equation for the reduction of acidified dichromate ions is shown below.

$Cr_2O_7^{2-}(aq) + 14H^+(aq) + 6e^- \longrightarrow 2Cr^{3+}(aq) + 7H_2O(l)$

(i) Write the half-equation for the oxidation of Fe^{2+} and construct the equation for the overall reaction with acidified dichromate (VI) ions.

$Fe^{2+}(aq) \longrightarrow Fe^{3+}(aq) + e^-$ ✓

$6Fe^{2+}(aq) + Cr_2O_7^{2-}(aq) + 14H^+(aq) \longrightarrow 6Fe^{3+}(aq) + 2Cr^{3+}(aq) + 7H_2O(l)$ ✓

Learn how to combine redox half-equations. Although this titration is different from the permanganate titrations that you have carried out, the same principles are used.
Even if your equation is wrong, you can still score in the next part provided that your method is built upon sound chemical principles. Show your working!

(ii) Using the information above, calculate the molar mass of the hydrated iron(II) sulfate and determine the number of moles of water of crystallisation present in one mole of hydrated iron(II) sulfate.

amount of $Cr_2O_7^{2-} = c \times \dfrac{V}{1000} = 0.00145 \times \dfrac{25.85}{1000} = 3.75 \times 10^{-5}$ mol ✓

From the equation, the amount of $Fe^{2+} = 6 \times$ amount of $Cr_2O_7^{2-}$. ✓

∴ amount of Fe^{2+} that reacted $= 6 \times 3.75 \times 10^{-5}$ mol $= 2.25 \times 10^{-4}$ mol ✓

In 250 cm³ solution, amount of $Fe^{2+} = 10 \times 2.25 \times 10^{-4} = 2.25 \times 10^{-3}$ mol ✓

∴ $\dfrac{0.626}{2.25 \times 10^{-3}} = 278$ g of hydrated iron(II) sulfate contains 1 mol Fe^{2+} ✓

M of hydrated iron(II) sulfate = 278 g mol⁻¹ ✓

M of $FeSO_4 = 55.8 + 32.1 + 16.0 \times 4 = 151.9$ g mol⁻¹ ✓

Number of waters of crystallisation $= \dfrac{278.0 - 151.9}{18.0} = \dfrac{126.1}{18.0} = 7$ ✓

Don't forget the scaling stage. The original solution was 250 cm³ but you have only taken 25 cm³ for the titration. Show your working!

(iii) Explain how the volume of potassium dichromate(VII) solution needed in the titration would be different if the solution of iron(II) sulfate had been left for some time before being titrated.

Less dichromate(VI) solution would have been needed ✓ because some of the Fe^{2+} would have been oxidised by air to Fe^{3+} ✓ [12]

[Total: 25]

The Periodic Table

Practice examination questions

1. (a) Complete the following table for the Period 3 oxides.

	Mg	Al	Si	P	S
Oxidation number	+2	+3	+4	+5	+6
Formula of oxide					

   [2]

   (b) Explain the trend in oxidation number. [1]

   (c) Write equations, if any, to show the effect of water on these oxides. [4]

   (d) What is the relationship between the structure and bonding of the Period 3 oxides and the pH of any resulting solution? [3]

   [Total: 10]

2. (a) Write the electron configuration of a nickel atom and a Ni^{2+} ion. [2]

   (b) Nickel is both a d-block element and a transition element. State what is meant by each of the terms. [2]

   (c) In aqueous solution, nickel sulfate, forms a six-coordinate complex ion which gives a green colour to the solution.

   (i) What feature of the water molecule allows it to form a complex ion with Ni^{2+}?

   (ii) Write the formula of the nickel complex ion formed in aqueous solution.

   (iii) What types of bond are present in this nickel complex ion?

   (iv) Suggest the shape of, and bond angles in, this complex ion. [6]

   (d) Consider the following reactions.

   (i) Write the formula of the green precipitate, **A**, and write an ionic equation for its formation above.

   (ii) Predict the formula of **B**.

   (iii) Write an equation for the formation of $[Ni(NH_3)_6]^{2+}$ from the aqueous nickel complex ion and suggest the type of reaction taking place. [5]

   [Total: 15]

The Periodic Table

Practice examination questions (continued)

3 (a) Deduce the oxidation state of the transition metal in each of the following species.
   (i) Cr in CrO_4^{2-}
   (ii) Cu in $[CuCl_2]^-$
   (iii) V in $[VO(H_2O)_5]^{2+}$ [3]

(b) Chromium(III) can be oxidised in alkaline conditions by hydrogen peroxide, H_2O_2. Use the half-equations below to construct the overall equation for this reaction.

$$Cr^{3+} + 8OH^- \longrightarrow CrO_4^{2-} + 4H_2O + 3e^-$$
$$H_2O_2 + 2e^- \longrightarrow 2OH^-$$ [1]

(c) Give two examples of the use of transition elements as catalysts in industrial processes. [2]

[Total: 6]

4 The concentration of an aqueous solution of hydrogen peroxide can be determined by titration. Aqueous potassium manganate(VII), $KMnO_4$, is titrated against a solution of hydrogen peroxide in the presence of acid.

The half-equations are shown below.

$$H_2O_2(aq) \longrightarrow 2H^+(aq) + O_2(g) + 2e^-$$
$$MnO_4^-(aq) + 8H^+(aq) + 5e^- \longrightarrow Mn^{2+}(aq) + 4H_2O(l)$$

(a) (i) Deduce the oxidation state of the manganese in the MnO_4^- ion.
   (ii) Construct the equation for the reaction between H_2O_2, MnO_4^- ions and H^+ ions. [2]

(b) State the role of hydrogen peroxide in this reaction. [1]

(c) After acidification, 25.0 cm³ of a dilute aqueous solution of hydrogen peroxide reacted exactly with 23.80 cm³ of 0.0150 mol dm⁻³ $KMnO_4$. Calculate the concentration, in mol dm⁻³, of hydrogen peroxide in the solution. [3]

[Total: 6]

5 Cobalt and nickel both form complex ions with the ligand ethane-1,2-diamine, $NH_2CH_2CH_2NH_2$.

(a) (i) Explain how ligands bond to a metal ion in a complex ion.
   (ii) What name is given to this type of ligand? [3]

(b) Cobalt forms the complex compound, $[Co(NH_2CH_2CH_2NH_2)_3]Cl_3$.
   (i) What is the oxidation state of cobalt in this compound?
   (ii) What is the electron configuration of cobalt in this compound? [2]

(c) In solution, nickel forms the complex ion $[Ni(NH_2CH_2CH_2NH_2)_3]^{2+}$.
   (i) Deduce the co-ordination number of nickel in this complex ion.
   (ii) State the shape around nickel in this complex ion.
   (iii) Outline how you could prepare a solution containing the complex ion $[Ni(NH_2CH_2CH_2NH_2)_3]^{2+}$ from an aqueous solution of nickel(II) chloride. Include an equation in your answer. [4]

[Total: 9]

Chapter 4
Chemistry of organic functional groups

The following topics are covered in this chapter:

- Isomerism and functional groups
- Aldehydes and ketones
- Carboxylic acids
- Esters
- Acylation
- Arenes
- Reactions of arenes
- Phenols and alcohols
- Amines

4.1 Isomerism and functional groups

After studying this section you should be able to:

- understand what is meant by structural isomerism and stereoisomerism
- explain the term chiral centre and identify any chiral centres in a molecule of given structural formula
- understand that many natural compounds are present as one optical isomer only
- understand that many pharmaceuticals are chiral drugs
- recognise common functional groups

Key points from AS
- Basic concepts
  Revise AS pages 96–101

During AS Chemistry, you learnt about the basic concepts used in organic chemistry. You should revise these thoroughly before you start the A2 part of this course.

Isomerism

AQA M4

Structural and *E/Z* isomerism were introduced during AS Chemistry. These are reviewed below and *E/Z* isomerism is discussed in the wider context of stereoisomerism.

Key points from AS
- Structural isomerism
  Revise AS pages 98–99
- E/Z isomerism
  Revise AS pages 106–107

Structural isomerism

Structural isomers are molecules with the same molecular formula but with different structural arrangements of atoms (structural formulae).

Two structural isomers of $C_2H_4O_2$ are shown below:

$$H_3C-C\begin{matrix}\nearrow O \\ \searrow OH\end{matrix} \qquad H-C\begin{matrix}\nearrow O \\ \searrow O-CH_3\end{matrix}$$

Stereoisomerism

Stereoisomers are molecules with the same structural formula but their atoms have different positions in space.

There are two types of stereoisomerism, each arising from a different structural feature:

- *E/Z* isomerism about a C=C double bond
- **optical** isomerism about a **chiral** carbon centre.

Chemistry of organic functional groups

E/Z isomerism

E/Z isomerism occurs in molecules with:

- a C=C double bond and
- two **different** groups attached to **each** carbon in the C=C bond.

The double bond prevents rotation.

E.g. *E/Z* isomers of 1,2-dichloroethene, ClCH=CHCl.

> In simple cases in which 2 of the groups are the same, *E/Z* isomerism is often referred to as *cis-trans* isomerism.

cis-isomer
(groups on one side)
Z isomer

trans-isomer
(groups on opposite sides)
E isomer

> Each *E/Z* isomer has the same structural formula.

Optical isomers

KEY POINT: Optical isomerism occurs in the molecules of a compound with a **chiral** (or *asymmetric*) carbon atom. A chiral carbon atom has **four** different groups attached to it.

E.g. the optical isomers of an amino acid, RCHNH$_2$COOH (see pages 116–117).

mirror plane

> Optical isomers are usually drawn as 3D diagrams. It is then easy to picture the mirror images.

Optical isomers:

- are non-superimposable mirror images of one another
- rotate plane-polarised light in opposite directions
- are chemically identical.

> Optical isomers are also called **enantiomers**.

Chirality and drug synthesis

A synthetic amino acid, made in the laboratory, is optically inactive and does not rotate plane-polarised light:

- it contains equal amounts of each optical isomer – a **racemic** mixture or **racemate**.

A natural amino acid, made by living systems, is optically active:

- it contains only one of the optical isomers.

> Society discovered the consequences of harmful side-effects from the 'wrong' optical isomer with the use of thalidomide. One optical isomer combated the effects of morning sickness in pregnant women. The other optical isomer was the cause of deformed limbs of unborn babies.
>
> Partly through this lesson, drugs are now often used as the optically pure form, comprising just the required optical isomer.

The difference between the optical isomers present in natural and synthetic organic molecules has important consequences for drug design. The synthesis of pharmaceuticals often requires the production of chiral drugs containing a single optical isomer. Although one of the optical isomers may have beneficial effects, the other may be harmful and may lead to undesirable side effects.

Progress check

1. Show the alkenes that are structural isomers of C$_4$H$_8$.
2. Show the *E/Z* isomers of C$_4$H$_8$.
3. Show the optical isomers of C$_4$H$_9$OH.

1 CH$_3$CH$_2$CH=CH$_2$; CH$_3$CH=CHCH$_3$; (CH$_3$)$_2$C=CH$_2$

Chemistry of organic functional groups

Common functional groups

AQA M4

The functional groups in the table below contain most of the functional groups you will meet in the A Level Chemistry course, including those met in AS Chemistry. It cannot be stressed enough just how important it is that you can instantly recognise a functional group. Without this, your progress in Organic Chemistry will be very limited.

It is essential that you can instantly identify a functional group within a molecule so that you can apply the relevant chemistry.

You must learn all of these!

Key points from AS
- Functional groups
  Revise AS pages 97–98

The nitrile carbon atom is included in the name. CH_3CN contains the longest carbon chain with **two** carbon atoms and its name is based upon ethane – hence ethanenitrile.

name	functional group	examples structural formula		prefix or suffix (for naming)
alkane	C–H	$CH_3CH_2CH_3$ propane	$CH_3CH_2CH_3$	-ane
alkene	$\diagdown$C=C$\diagdown$	CH_3CHCH_2 propene	H_3C \ C=C / H / \ H H	-ene
halogenoalkane	—Br	CH_3CH_2Br bromoethane	CH_3CH_2—Br	bromo-
alcohol	—OH	CH_3CH_2OH ethanol	CH_3CH_2—OH	-ol
aldehyde	—C(=O)H	CH_3CHO ethanal	H_3C—C(=O)H	-al
ketone	—C(=O)—	CH_3COCH_3 propanone	H_3C—C(=O)—CH_3	-one
carboxylic acid	—C(=O)OH	CH_3COOH ethanoic acid	H_3C—C(=O)OH	-oic acid
ester	—C(=O)O—	CH_3COOCH_3 methyl ethanoate	H_3C—C(=O)O—CH_3	-oate
acyl chloride	—C(=O)Cl	CH_3COCl ethanoyl chloride	H_3C—C(=O)Cl	-oyl chloride
amine	—NH_2	$CH_3CH_2NH_2$ ethylamine	CH_3CH_2—NH_2	-amine
amide	—C(=O)NH_2	CH_3CONH_2 ethanamide	H_3C—C(=O)NH_2	-amide
nitrile	—C≡N	CH_3CN ethanenitrile	H_3C—CN	-nitrile

Chemistry of organic functional groups

4.2 Aldehydes and ketones

After studying this section you should be able to:

- understand the polarity and physical properties of carbonyl compounds
- describe the reduction of carbonyl compounds to form alcohols
- describe nucleophilic addition to aldehydes and ketones
- describe tests to detect the presence of an aldehyde group

LEARNING SUMMARY

Carbonyl compounds General formula: $C_nH_{2n}O$

During AS Chemistry, you learnt about how alcohols can be oxidised to carbonyl compounds: aldehydes and ketones. For A2 Chemistry, you will learn about the reactions of aldehydes and ketones.

Key points from AS

- Oxidation of alcohols
  Revise AS pages 112–113

Types and naming of carbonyl compounds

The carbonyl group, C=O, is the functional group in aldehydes and ketones.

aldehyde, RCHO

ketone, RCOR'

ethanal
CH_3CHO

butanal
$CH_3CH_2CH_2CHO$

propanone
CH_3COCH_3

pentan-2-one
$CH_3COCH_2CH_2CH_3$

Polarity of carbonyl compounds

Carbon and oxygen have different electronegativities, resulting in a polar C=O bond: carbonyl compounds have polar molecules.

> The properties of aldehydes and ketones are dominated by the polar carbonyl group, C=O.

oxygen is more electronegative than carbon producing a dipole

Physical properties of carbonyl compounds

> The polarity in propanone is such that it mixes with polar solvents such as water and also dissolves many organic compounds.
> The low boiling point also makes it easy to remove by evaporation, a property exploited by its use in paints and varnishes.

The polarity of the carbonyl group is less than that of the hydroxyl group in alcohols. Thus, aldehydes and ketones have weaker dipole–dipole interactions and lower boiling points than alcohols of comparable molecular mass.

carbonyl group in aldehyde or ketone → stronger dipole → hydroxyl group in alcohols

Reduction of carbonyl compounds

Aldehydes and ketones can be reduced to alcohols using a reducing agent containing the hydride ion, H^-.

Suitable reducing agents are:
- sodium tetrahydridoborate(III) (*sodium borohydride*), $NaBH_4$, in water (heat)
- lithium tetrahydridoaluminate(III) (*lithium aluminium hydride*), $LiAlH_4$, in ether.

Chemistry of organic functional groups

Aldehydes are reduced to primary alcohols:

For balanced equations, the reducing agent can be shown simply as [H].

$$H_3C-CHO + 2[H] \longrightarrow H_3C-CH_2OH$$
aldehyde → primary alcohol

NaBH$_4$ and LiAlH$_4$ both reduce the C=O double bond in aldehydes and ketones.

They do **not** reduce the C=C bond in alkenes.

Ketones are reduced to secondary alcohols:

$$\begin{array}{c}H_3C\\C_2H_5\end{array}C=O + 2[H] \longrightarrow \begin{array}{c}H_3C\\C_2H_5\end{array}CHOH$$
ketone → secondary alcohol

Nucleophilic addition to carbonyl compounds

The electron-deficient carbon atom of the polar $C^{\delta+}=O^{\delta-}$ bond attracts nucleophiles. This allows an addition reaction to take place across the C=O double bond of aldehydes and ketones. This is called **nucleophilic addition**.

$$\overset{\delta+}{C}=\overset{\delta-}{O} \quad \text{electron-rich nucleophile attracted to } \overset{\delta+}{C}$$

Reduction as nucleophilic addition

The reaction of carbonyl compounds with hydrogen cyanide is also an example of nucleophilic addition.

The reduction of aldehydes and ketones to alcohols using NaBH$_4$ or LiAlH$_4$ is an example of **nucleophilic addition**. The reducing agent can be considered to release hydride ions, H$^-$.

$$NaBH_4 \longrightarrow H^- + NaBH_3^+$$

The hydride ion acts as a nucleophile in the first stage of the reaction.

Mechanism

$$H_3C-CHO \xrightarrow{:H^-} H_3C-C(H)(H)-O^- \xrightarrow{\text{protonation with } H_2O} H_3C-C(H)(H)-OH + OH^-$$

:H$^-$ nucleophile donates electron pair → primary alcohol

Nucleophilic addition of hydrogen cyanide, HCN

HCN is added across the C=O double bond.

In the presence of cyanide ions, CN$^-$, hydrogen cyanide, HCN, is added across the C=O bond in aldehydes and ketones.

aldehyde + HCN ⟶ hydroxynitrile
CH$_3$CHO + HCN ⟶ CH$_3$CH(OH)CN

$$H_3C-CHO \xrightarrow{KCN(aq) / HCN} H_3C-C(H)(OH)-C\equiv N$$

This hydroxynitrile has optical isomers.

Hydrogen cyanide is a very poisonous gas and it is usually generated in solution as H$^+$ and CN$^-$ ions using:
- sodium cyanide, NaCN as a source of CN$^-$
- dilute sulfuric acid as a source of H$^+$.

Chemistry of organic functional groups

> The presence of cyanide ions is essential to provide the nucleophile for the first step of this mechanism.

Mechanism

$$H_3C-\overset{\delta+}{C}\overset{\delta-}{\underset{H}{\overset{O}{\diagup}}} \quad :C\equiv N \longrightarrow H_3C-\underset{H}{\overset{:\ddot{O}:^-}{\underset{|}{C}}}-C\equiv N \xrightarrow{H^+ \text{ transfer}} H_3C-\underset{H}{\overset{O-H}{\underset{|}{C}}}-C\equiv N$$

:CN nucleophile donates electron pair · 'hydroxynitrile'

Increasing the carbon chain length

The nucleophilic addition of hydrogen cyanide is useful for increasing the length of a carbon chain. The nitrile product can then easily be reacted further in organic synthesis.

- Nitriles are easily *hydrolysed* by water in hot dilute acid to form a carboxylic acid.
- Nitriles are easily *reduced* by sodium in ethanol to form an amine.

The diagrams below show how 2-hydroxypropanenitrile, synthesised above, can be converted into a carboxylic acid and an amine.

nitrile · *carboxylic acid*

$$H_3C-\underset{H}{\overset{OH}{\underset{|}{C}}}-C\equiv N \;+\; 2H_2O \;+\; H^+ \xrightarrow[\text{hydrolysis}]{\text{aqueous acid reflux}} H_3C-\underset{H}{\overset{OH}{\underset{|}{C}}}-C\overset{O}{\underset{OH}{\diagdown}} \;+\; NH_4^+$$

2-hydroxypropanoic acid (lactic acid)

$$H_3C-\underset{H}{\overset{OH}{\underset{|}{C}}}-C\equiv N \;+\; 2[H] \xrightarrow[\text{reduction}]{\text{Na/ethanol OR LiAlH}_4} H_3C-\underset{H}{\overset{OH}{\underset{|}{C}}}-\underset{H}{\overset{H}{\underset{|}{C}}}-NH_2$$

nitrile · *amine*

Testing for the aldehyde group

AQA M4

The oxidation of aldehydes provides the basis of chemical tests used to identify this functional group.

> An aldehyde can be distinguished from a ketone by using a combination of these two tests.
>
> Both aldehyde and ketone produce a yellow–orange precipitate with 2,4-DNPH. Only the aldehydes react in the three tests below.
>
> Ketones cannot normally be oxidised and they **do not** react with Tollens' reagent, Fehling's solution or acidified dichromate(VI).
>
> Tollens' reagent contains the [Ag(NH$_3$)$_2$] complex ion (see page 67).

KEY POINT

In each test:
- the aldehyde **reduces** the reagents used in the test, producing a visible colour change
- the aldehyde is **oxidised** to a carboxylic acid.
  RCHO + [O] ⟶ RCOOH

Heat aldehyde with Tollens' reagent

Tollens' reagent is a solution of silver nitrate in aqueous ammonia. The silver ions are reduced by an aldehyde producing a **silver mirror**.

KEY POINT

Tollens' reagent ⟶ silver mirror
Ag$^+$(aq) + e$^-$ ⟶ Ag(s)

Chemistry of organic functional groups

Heat aldehyde with Fehling's solution

Fehling's solution contains Cu^{2+} ions dissolved in aqueous alkali. The Cu^{2+} ions are reduced to copper(I) ions, Cu^+, producing a **brick-red precipitate** of Cu_2O.

> Fehling's solution ⟶ **brick-red precipitate** of Cu_2O
>
> $2Cu^{2+}(aq) + 2e^- + 2OH^-(aq) \longrightarrow Cu_2O(s) + H_2O(l)$

Key points from AS

- Oxidation of primary alcohols
  Revise AS pages 112–113

Acidified dichromate(VI) can also be used to oxidise alcohols.

Heat aldehyde with $H^+/Cr_2O_7^{2-}$

Concentrated sulfuric acid, H_2SO_4, is used as a source of H^+ ions and potassium dichromate(VI), $K_2Cr_2O_7$, as a source of $Cr_2O_7^{2-}$ ions. The **orange** $Cr_2O_7^{2-}$ ions are reduced to **green** chromium(III) ions, Cr^{3+}.

Progress check

1. (a) Draw the structure of:
   - (i) 3-methylpentan-2-one
   - (ii) 3-ethyl-2-methylpentanal.

2. **Three** isomers of C_4H_8O are carbonyl compounds.
   (a) (i) Show the formula for each isomer.
   (ii) Classify each isomer as an *aldehyde* or a *ketone*.
   (iii) Name each isomer.
   (b) Write the structural formulae of the **three** alcohols that could be formed by reduction of these isomers.

1 (a) (i) $H_3C-CH_2-\underset{O}{\overset{\|}{C}}-\underset{CH_3}{\overset{|}{CH}}-CH_3$

(ii) $H_3C-CH-CH_2-CH-\underset{\underset{CH_2CH_3}{|}}{C}\overset{H}{\underset{\|O}{}}$ with CH_3 branch

2 (a) (i)
$CH_3-CH_2-CH_2-C\overset{H}{\underset{\|O}{}}$ butanal aldehyde (ii)

$CH_3-CH_2-\underset{O}{\overset{\|}{C}}-CH_3$ butanone ketone

$CH_3-\underset{CH_3}{\overset{|}{CH}}-C\overset{H}{\underset{\|O}{}}$ methylpropanal aldehyde

(b) $CH_3CH_2CH_2CH_2OH$; $CH_3CH_2CHOHCH_3$; $(CH_3)_2CHCH_2OH$

Chemistry of organic functional groups

4.3 Carboxylic acids

LEARNING SUMMARY

After studying this section you should be able to:

- understand the polarity and physical properties of carboxylic acids
- describe acid reactions of carboxylic acids to form salts
- describe the esterification of carboxylic acids with alcohols

Carboxylic acids   General formula: $C_nH_{2n+1}COOH$ RCOOH

> The combination of carbonyl and hydroxyl groups in the carboxyl group modifies the chemistry of both groups.
> Carboxylic acids have their own set of reactions and react differently from carbonyl compounds and alcohols.

The carboxyl group

The functional group in carboxylic acids is the **carboxyl** group, COOH. Although this combines both the **carbonyl** group and a **hydroxyl** group its properties are very different from either.

Naming of carboxylic acids

Numbering starts from the carbon atom of the carboxyl group.

4-methylpentanoic acid
$(CH_3)_2CH_2CH_2CH_2COOH$

Natural carboxylic acids

Carboxylic acids are found commonly in nature. Their acidity is comparatively weak and their presence in food often gives a sour taste. Some examples of natural carboxylic acids are shown below.

structure	name	natural source
HCOOH	methanoic acid (*formic acid*)	ants, stinging nettles
CH_3COOH	ethanoic acid (*acetic acid*)	vinegar
COOH | COOH	ethanedioic acid (*oxalic acid*)	rhubarb
OH | H_3C—CH—COOH	2-hydroxypropanoic acid (*lactic acid*)	sour milk
CH_2COOH | HO—C—COOH | CH_2COOH	2-hydroxypropane-1,2,3-tricarboxylic acid (*citric acid*)	oranges, lemons

Polarity

The carboxyl group is a combination of two polar groups: the hydroxyl –OH, **and** carbonyl C=O groups. This makes a carboxylic acid molecule more polar than a molecule of an alcohol or a carbonyl compound.

carbonyl group in aldehyde or ketone → stronger dipole → hydroxyl group in alcohols → stronger dipole → carboxyl group in carboxylic acids

Chemistry of organic functional groups

The properties of carboxylic acids are dominated by the carboxyl group, COOH.

Physical properties of carboxylic acids

The COOH group dominates the physical properties of short-chain carboxylic acids. Hydrogen bonding takes place between carboxylic acid molecules, resulting in:

- higher melting and boiling points than alkanes of comparable M_r
- solubility in water.

The solubility of alcohols in water decreases with increasing carbon chain length as the non-polar contribution to the molecule becomes more important.

Carboxylic acids as 'acids'

AQA M4

Carboxylic acids are only weak acids because they only partially dissociate in water.

$$CH_3COOH \rightleftharpoons CH_3COO^- + H^+$$

- Only 1 molecule in about 100 actually dissociates.
- Only a small proportion of the potential H^+ ions is released.

Carboxylic acids are the 'organic acids'.

For more details of the dissociation of weak acids, see pages 25–26.

Carboxylates: salts of carboxylic acids

Carboxylic acid salts, 'carboxylates', are formed by neutralisation of a carboxylic acid by an alkali. In the example below, ethanoic acid produces ethanoate ions.

carboxylic acids exist in acidic conditions

carboxylates exist in alkaline conditions

On evaporation of water, a carboxylate salt crystallises out as an ionic compound. With aqueous sodium hydroxide as the alkali, sodium ethanoate, $CH_3COO^-Na^+$, is produced as the ionic salt.

Acid reactions of carboxylic acids

Carboxylic acids take part in typical acid reactions. Note in the examples below that each salt formed is a carboxylate.

- They are **neutralised by alkalis**, forming a salt and water only.
  $$CH_3COOH + NaOH \longrightarrow CH_3COONa + H_2O$$
- They **react with carbonates**, forming a salt, carbon dioxide and water.
  $$2CH_3COOH + CaCO_3 \longrightarrow (CH_3COO)_2Ca + CO_2 + H_2O$$
- They **react with** reactive **metals**, forming a salt and hydrogen.
  $$2CH_3COOH + Mg \longrightarrow (CH_3COO)_2Mg + H_2$$

Carboxylic acids react by the usual 'acid reactions' producing carboxylate salts.

> **KEY POINT**
> Carboxylic acids are the only common organic group able to release carbon dioxide gas from carbonates. This provides a useful test to show the presence of a carboxyl group.

Properties of carboxylates

Carboxylates such as sodium ethanoate, $CH_3COO^-Na^+$, are ionic compounds. They have typical properties of an ionic compound.

- They are solids at room temperature with high melting and boiling points.
- They have a giant ionic lattice structure.
- They dissolve in water, totally dissociating into ions.

Chemistry of organic functional groups

Esterification

AQA M4

Esterification is the formation of an ester by reaction of a **carboxylic acid** with an **alcohol** in the presence of an **acid catalyst** (e.g. concentrated sulfuric acid).

carboxylic acid + alcohol ⟶ ester + water

The esterification of ethanoic acid by methanol is shown below:

$$CH_3COOH + CH_3OH \longrightarrow CH_3COOCH_3 + H_2O$$

> Esters are formed by reaction of a carboxylic acid with an alcohol.

Conditions

- An acid catalyst (a few drops of conc. H_2SO_4) and reflux.
- The yield is usually poor due to incomplete reaction.

Progress check

1. Write down the structural formula of:
   (a) propanoic acid
   (b) the propanoate ion.

2. Explain why a carboxylic acid has a higher boiling point than the corresponding alcohol.

1 (a) CH_3CH_2COOH
  (b) $CH_3CH_2COO^-$
2 Carboxylic acid has both polar carbonyl and hydroxyl groups. Alcohols have hydroxyl group only. Therefore, a carboxylic acid has greater intermolecular forces.

Chemistry of organic functional groups

4.4 Esters

After studying this section you should be able to:

- understand the physical properties of esters
- describe the acid and base hydrolysis of esters
- understand that fats and oils are saturated and unsaturated esters
- describe the hydrolysis of fats and oils in soap making

Esters

General formula: $C_nH_{2n+1}COOC_mH_{2m+1}$ RCOOR'

AQA M4

The functional group in esters is the COOR' group, with an alkyl group in place of the acidic proton of a carboxylic acid.

Naming of esters

The name of an ester is based on the carboxylic acid from which the ester is derived. The ester below, methyl butanoate, is derived from butanoic acid C_3H_7COOH with a methyl group in place of the acidic proton.

butanoate from
butanoic acid
$CH_3CH_2CH_2COOH$

$CH_3CH_2CH_2-C\overset{O}{\underset{O-CH_3}{\diagdown\!\!/}}$

methyl from methanol
CH_3OH

methyl butanoate

In the name of **methyl butanoate**:

- the alkyl group comes first as the ...*yl*: **methyl**
- the carboxylic acid part comes second as the ...*oate*: **butanoate**

Natural esters

Esters are found commonly in nature as fats and oils. They often have pleasant smells and contribute to the flavouring of many foods as shown below.

Esters are common organic compounds present in fats and oils. They are also used as solvents, plasticisers, in food flavourings and as perfumes.

structure	name	source
$HCOOCH_3$	methyl methanoate	raspberries
$C_3H_7COOC_4H_9$	butyl butanoate	pineapple

Physical properties of esters

Unlike carboxylic acids, esters are neutral. They are less polar than carboxylic acids. The absence of an –OH group means that esters cannot form hydrogen bonds and are generally insoluble in water.

Hydrolysis of esters

AQA M4

The hydrolysis of an ester is the reverse reaction to esterification (see page 96):

hydrolysis
ester + water ⇌ carboxylic acid + alcohol
esterification

Note that this reaction is reversible. The direction of reaction can be controlled by the reagents and reaction conditions used.

> **KEY POINT**
> Hydrolysis is the breaking down of a compound using **water** as the reagent.

97

Chemistry of organic functional groups

Hydrolysis takes place by refluxing the ester with dilute aqueous acid or alkali.
- Acid hydrolysis ⟶ alcohol + carboxylic acid.
- Alkaline hydrolysis ⟶ alcohol + carboxylate.

*Esterification **produces** water.*
*Hydrolysis **reacts** with water.*

Fats and oils

AQA M4

Natural fats and oils are *triglyceryl esters* of fatty acids (long chain carboxylic acids) and propane-1,2,3-triol (*glycerol*).

```
RCOO—CH₂            R—COOH           HO—CH₂
    |                                     |
RCOO—CH             R—COOH           HO—CH
    |                                     |
RCOO—CH₂            R—COOH           HO—CH₂
 fat or oil         fatty acids         glycerol
triglyceryl ester
```

Most fats and oils are mixed esters from different fatty acids. So 'R' in these structures may refer to three different chains.

Saturated and unsaturated fats

Saturated fats tend to be solids at room temperature and are mainly from animal products. Unsaturated fats tend to be oils and are mainly from vegetable products.

Saturated and unsaturated fatty acids can be obtained from fats and oils by hydrolysis (see page 99). An unsaturated fatty acid would have at least one double C=C bond in one of the alkyl carbon chains.

Examples of saturated and unsaturated fatty acids

octadecanoic acid, 18,0

octadec-9-enoic acid, 18,1(9);

octadeca-9,12-dienoic acid, 18,2(9,12)

- Note that the unsaturated fatty acids above are *trans* acids. The *cis* isomers are less compact and are far healthier.
- Many triglyceryl esters contain different alkyl carbon chains derived from different fatty acids.

Scientists have discovered that trans fatty acids pose health problems, increasing cholesterol with an increased risk of coronary heart disease, strokes and obesity.

In margarine production, vegetable oils are hardened by partial **catalytic hydrogenation**.

- Partial hydrogenation leaves some double bonds intact. The oil is reacted just enough to solidify the oily texture. This prevents the formation of **polysaturates**, with no C=C double bonds, which are linked to unhealthy diets.
- Margarines are often sold as partially hydrogenated **polyunsaturates** to promote the health value of unsaturated fats.

Chemistry of organic functional groups

Hydrolysis of fats and oils

The hydrolysis of fats and oils is an important reaction used in soap production in which the fat or oil is boiled with aqueous alkali. In alkaline hydrolysis with aqueous NaOH, each mole of the fat or oil produces:

- **three** moles of the fatty acid salt, $3RCOO^-Na^+$ and
- **one** mole of glycerol (a triol), $HOCH_2CHOHCH_2OH$.

$$\begin{array}{c} RCOO-CH_2 \\ | \\ RCOO-CH \\ | \\ RCOO-CH_2 \end{array} + 3\ NaOH \xrightarrow{NaOH(aq)} 3\ RCOO^-Na^+ \text{ (salt of carboxylic acid)} + \begin{array}{c} HO-CH_2 \\ | \\ HO-CH \\ | \\ HO-CH_2 \end{array}$$

Biodiesel

AQA M4

Transesterification is also being investigated for the development of low-fat spreads as an alternative to hydrogenation of vegetable oils to produce margarine.

Esters of fatty acids are being developed for use as fuels such as biodiesel. Use of biodiesel as a fuel increases the contribution to energy requirements from renewable fuels. Biodiesel is a mixture of esters of long chain carboxylic acids.

Vegetable oils can be converted into biodiesel by **transesterification** using methanol in the presence of an alkaline catalyst, e.g.

$$\begin{array}{c} RCOO-CH_2 \\ | \\ RCOO-CH \\ | \\ RCOO-CH_2 \end{array} + 3\ CH_3OH \xrightarrow[\text{catalyst}]{NaOH} 3\ RCOO-CH_3 \text{ (methyl ester)} + \begin{array}{c} HO-CH_2 \\ | \\ HO-CH \\ | \\ HO-CH_2 \end{array}$$

The methyl ester is the easiest to prepare but methanol has the disadvantage that it is toxic. Research is being done to develop the ethyl ester made from ethanol. Ethanol is easy to produce from fermentation of sugars.

Progress check

1. How is methyl propanoate made from a named carboxylic acid and alcohol?
2. (a) Explain what is meant by the term *hydrolysis*.
   (b) Name the products of the acid and alkaline hydrolysis of ethyl butanoate.

1 Reflux propanoic acid with methanol in the presence of a few drops of concentrated sulfuric acid as catalyst.
2 (a) Breaking down a molecule using water.
(b) Acid hydrolysis: ethanol and butanoic acid.
Alkaline hydrolysis: ethanol and butanoate ions.

Chemistry of organic functional groups

4.5 Acylation

After studying this section you should be able to:

- describe the formation of acyl chlorides from carboxylic acids
- describe nucleophilic addition-elimination reactions of acyl chlorides
- know the advantages of ethanoic anhydride for industrial acylations

Acyl chlorides

General formula: $C_nH_{2n+1}COCl$ RCOCl

The functional group in acyl chlorides is the COCl group:

$$-C\begin{smallmatrix}\diagup O \\ \diagdown Cl\end{smallmatrix}$$

Acyl chlorides are named from the parent carboxylic acid.

The examples below show that the suffix '*-oic acid*' is changed to '*-oyl chloride*' in the corresponding acyl chloride.

| ethanoic acid | ethanoyl chloride | propanoic acid | propanoyl chloride |

$H_3C-C(=O)OH$ $H_3C-C(=O)Cl$ $CH_3CH_2-C(=O)OH$ $CH_3CH_2-C(=O)Cl$

Properties

Acyl chlorides are the most reactive organic compounds that are commonly used. Unlike most other organic functional groups, acyl chlorides do **not** occur naturally because of their high reactivity with water.

> Acyl chlorides are the only common organic compounds that react violently with water.

$R-C\begin{smallmatrix}\delta+ \diagup O^{\delta-} \\ \diagdown Cl_{\delta-}\end{smallmatrix}$ The relatively large $\delta+$ charge attracts the lone pair of a nucleophile

The electron-withdrawing effects of both the **chlorine** atom and carbonyl **oxygen** atom produces a relatively large $\delta+$ charge on the carbonyl carbon atom. Nucleophiles are strongly attracted to the electron-deficient carbon atom, increasing the reactivity.

Acyl chlorides in organic synthesis

The high reactivity of acyl chlorides compared with carboxylic acids makes them particularly useful in the organic synthesis of related compounds.

Advantages of acyl chlorides over carboxylic acids:

- A good yield of product – reactions go to completion.
- Reactions often occur quickly and at lower temperatures.

Reactions of acyl chlorides with nucleophiles

Nucleophiles of the type H–Y : react readily with acyl chlorides.

lone pair attracted to $C^{\delta+}$

$H_3C-C\begin{smallmatrix}\delta+ \diagup O^{\delta-} \\ \diagdown Cl_{\delta-}\end{smallmatrix}$:Y$^{\delta-}$ nucleophile
       H $^{\delta+}$
HCl forms

$\longrightarrow$ $H-\underset{H}{\overset{H}{C}}-C\begin{smallmatrix}\diagup O \\ \diagdown Y\end{smallmatrix}$ + H–Cl

Chemistry of organic functional groups

This is an **addition-elimination** reaction involving:
- addition of HY across the C=O double bond followed by
- elimination of HCl.

Addition-elimination reactions of acyl chlorides

Acyl chlorides can be reacted with different nucleophiles to produce a range of related functional groups. In each addition-elimination reaction, hydrogen chloride is eliminated as the second product.

The reaction scheme below shows the reactions of an acyl chloride RCOCl with the nucleophiles water, H_2O, methanol, CH_3OH, ammonia, NH_3 and methylamine, CH_3NH_2.

The high reactivity of an acyl chloride means that these reactions take place at room temperature.

Ammonia and methylamine are bases – they react with the HCl eliminated:

$$NH_3 + HCl \longrightarrow NH_4^+Cl^-$$
$$CH_3NH_2 + HCl \longrightarrow CH_3NH_3^+Cl^-$$

Mechanism

The mechanism below shows the **nucleophilic addition-elimination** reaction of ethanoyl chloride with methanol to form an ester. Each nucleophile in the reaction scheme above will react in a similar manner.

Acid anhydrides

AQA — M4

Acyl chlorides are ideal for small-scale preparations in the laboratory. However, they are too expensive and too reactive for large-scale preparations, for which they are replaced by acid anhydrides.

Chemistry of organic functional groups

The equation below shows the reaction between ethanoic anhydride and methanol:

> Notice the molecular structure of ethanoic anhydride – two ethanoic acid molecules have been condensed together with loss of H_2O. Hence the 'anhydride' from ethanoic acid.

acid anhydride → carboxylic acid as second product

- The reaction is slower and easier to control on a large scale.
- This is the same essential reaction as with an acyl chloride but RCOOH forms instead of HCl as the second product.

Synthesis of aspirin

Aspirin can be synthesised by acylation of 2-hydroxybenzenecarboxylic acid (*salicylic acid*) using ethanoic anhydride.

→ aspirin + ethanoic acid as second product

Progress check

1. Write down the structural formula of the organic product formed from the reaction of propanoyl chloride with:
   (a) water
   (b) ammonia.
2. What could you react together to make N-ethylbutanamide, $CH_3CH_2CH_2CONHCH_2CH_3$?
3. Write an equation for the preparation of aspirin from 2-hydroxybenzoic acid using an acyl chloride.

1 (a) CH_3CH_2COOH
  (b) $CH_3CH_2CONH_2$
2 $CH_3CH_2CH_2COCl$ and $CH_3CH_2NH_2$
3

Chemistry of organic functional groups

4.6 Arenes

LEARNING SUMMARY

After studying this section you should be able to:

- apply rules for naming simple aromatic compounds
- understand the delocalised model of benzene
- explain the resistance to addition of benzene compared with alkenes

Aromatic organic compounds

AQA   M4

Organic compounds with pleasant smells were originally classified as aromatic compounds. Many of these contain a benzene ring in their structure and nowadays an aromatic compound is one structurally derived from benzene, C_6H_6.

The diagrams below show different representations of a benzene molecule. It is usual practice to omit the carbon and hydrogen labels.

Arenes burn with a smoky flame. This reflects the relatively low hydrogen to carbon ratio compared to alkanes and alkenes.

Arenes

An **arene** is an aromatic hydrocarbon containing a benzene ring. Benzene is the simplest arene and substituted arenes have alkyl groups attached to the benzene ring. Examples of arenes are shown below.

benzene   methylbenzene   ethylbenzene

Functional groups

Arenes can have a functional group next to an **aryl** group (a group containing a benzene ring):

Aryl group   Functional group

or C_6H_5X

An aryl group contains a benzene ring.

- An aryl group is often represented simply as Ar—.
- The simplest aryl group is the **phenyl** group, C_6H_5, derived from benzene, C_6H_6.

Naming of aromatic organic compounds

The benzene ring of an aromatic compound is numbered from the carbon atom attached to a functional group or alkyl side-chain.
The names can be derived in two ways:

103

Chemistry of organic functional groups

- Some compounds (e.g. hydrocarbons, chloroarenes and nitroarenes) are regarded as substituted benzene rings.

chlorobenzene   1,3-dimethylbenzene   2,4-dinitromethylbenzene

You should be able to suggest names for simple arenes.

- Other compounds (e.g. phenols and amines) are considered as phenyl compounds of a functional group.

phenol
not hydroxybenzene

phenylamine
not aminobenzene

2,4,6-trichlorophenol

The stability of benzene

AQA M4

Two structures are used to represent benzene:

- the **Kekulé** structure, developed between 1865 and 1872
- the modern **delocalised**, structure developed in the 1930s.

The Kekulé structure of benzene

The original Kekulé structure showed benzene as a hexagonal molecule with alternate double and single bonds. Each carbon atom is attached to one hydrogen atom. This model was later modified to one with two isomers, rapidly interconverting into one another.

The Kekulé model of a benzene molecule has alternate double and single bonds making up the ring.

The Kekulé structure of benzene

*The chemical name for the Kekulé structure of benzene is **cyclohexa-1,3,5-triene**, after the positions of the double bonds in the ring.*

The delocalised structure of benzene

The delocalised structure of benzene shows a benzene molecule as a hybrid state between Kekulé's two isomers with no separate single and double bonds.

In this hybrid state:

The delocalised model of a benzene molecule has identical carbon–carbon bonds making up the ring.

- each carbon atom contributes one electron from its p-orbital to form π-bonds
- the π-bonds are spread out or **delocalised** over the whole ring.

The double bond in an alkene is a **localised** π-bond between two carbon atoms. In the delocalised structure of benzene, each bond is identical, with electron density spread out to encompass and stabilise the whole ring. The diagram below shows one of the π-bonds formed by delocalisation of electrons in a benzene molecule.

Key point from AS

- **Alkenes**
  Revise AS page 106

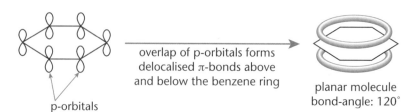

p-orbitals

overlap of p-orbitals forms
delocalised π-bonds above
and below the benzene ring

planar molecule
bond-angle: 120°

Chemistry of organic functional groups

Although the Kekulé structure is used for some purposes, the delocalised structure is a better representation of benzene. You will find both representations in books.

The delocalised structure of a benzene molecule is shown as a hexagon to represent the carbon skeleton and a circle to represent the six delocalised electrons.

delocalised electrons
– all carbon-carbon bonds identical

The shape of a benzene molecule

Key points from AS

- Electron-pair repulsion theory
  Revise AS pages 45–46

Using electron-pair repulsion theory:

- there are **three** centres of electron density surrounding each carbon atom
- the shape around each carbon atom is trigonal planar with bond angles of 120°.

This results in a benzene molecule that is **planar**.

3 electron centres surround each carbon atom

Experimental evidence for delocalisation

Bond length data

The Kekulé structure of benzene as cyclohexa-1,3,5-triene suggests two bond lengths for the separate single and double bonds:

- C—C bond length = 0.154 nm
- C=C bond length = 0.134 nm

Experiment shows only one C—C bond length of 0.139 nm, between the bond lengths for single and double carbon-carbon bonds.

0.154 nm   0.134 nm
0.134 nm   0.154 nm
0.154 nm   0.134 nm

all C–C bonds are the same length: 0.139 nm

> This shows that each carbon-carbon bond in the benzene ring is intermediate between a single and a double bond.

Thermochemical evidence

Hydrogenation of cyclohexene

Each molecule of cyclohexene has **one** C=C double bond. The enthalpy change for the reaction of cyclohexene with hydrogen is shown below:

$$\text{cyclohexene} + H_2 \longrightarrow \text{cyclohexane} \qquad \Delta H^\ominus = -120 \text{ kJ mol}^{-1}$$

105

Chemistry of organic functional groups

Hydrogenation of benzene

The Kekulé structure of benzene as cyclohexa-1,3,5-triene has **three** double C=C bonds. It would be expected that the enthalpy change for the hydrogenation of this structure would be three times the enthalpy change for the **one** C=C bond in cyclohexene.

The stability of benzene can also be demonstrated using thermochemical data for the reactions of bromine with cyclohexene and benzene.

predicted enthalpy change:
$\Delta H^\ominus = 3 \times -120 = -360$ kJ mol^{-1}

- When benzene is reacted with hydrogen, the enthalpy change obtained is far less exothermic, $\Delta H^\ominus = -208$ kJ mol^{-1}.

> **KEY POINT**
> - The difference between the thermochemical data for cyclohexa-1,3,5-triene and benzene suggests that benzene has **more stable bonding** than the Kekulé structure.
> - The delocalisation or resonance energy of benzene of -152 kJ mol^{-1} is the difference between the two enthalpy changes above. This is extra energy that must be provided to break the delocalised benzene ring.

Progress check

1. Name the **three** aromatic isomers of $C_6H_4Br_2$.
2. State two pieces of experimental evidence that support the delocalised structure of benzene.
3. Explain why alkenes, such as cyclohexene, are so much more reactive with electrophiles than arenes such as benzene.

1. 1,2-dibromobenzene; 1,3-dibromobenzene; 1,4-dibromobenzene.
2. The enthalpy change of hydrogenation of benzene is less exothermic than three times the enthalpy change of hydrogenation of cyclohexene. All the carbon-carbon bonds in the benzene ring are the same length in between a double and single bond.
3. Alkenes have localised π-bonds with a larger electron density than the delocalised π-bonds in benzene. The greater electron density is able to attract electrophiles more strongly. Also, the stability of the benzene ring must be disrupted if benzene is to react.

Chemistry of organic functional groups

4.7 Reactions of arenes

LEARNING SUMMARY

After studying this section you should be able to:
- describe electrophilic substitution of arenes: nitration, alkylation and acylation
- describe the mechanism of electrophilic substitution in arenes
- understand the importance of reactions of arenes in the synthesis of commercially important materials

Electrophilic substitution reactions of arenes

AQA M4

Many electrophiles react with alkenes by **addition**. However, electrophiles react with arenes by **substitution**, replacing a hydrogen atom on the ring. The difference in behaviour results from the high stability of the **delocalised** π-bonds in benzene compared with the **localised** π-bonds in alkenes (see pages 104–105).

KEY POINT
- Benzene reacts with only very reactive electrophiles.
- The typical reaction of an arene is **electrophilic substitution**.

In this section, reactions of benzene are discussed to illustrate electrophilic substitution reactions of arenes.

Nitration of arenes

AQA M4

The nitration of arenes produces aromatic nitro compounds, which are important for the synthesis of many important products including explosives and dyes.

Nitration of benzene

Benzene is nitrated by concentrated nitric acid at 55°C in the presence of concentrated sulfuric acid, which acts as a catalyst.

Although some heat is required (55°C), too much may give further nitration of the benzene ring forming 1,3-dinitrobenzene.

C$_6$H$_6$ + HNO$_3$ $\xrightarrow{\text{H}_2\text{SO}_4, 55°C}$ C$_6$H$_5$NO$_2$ + H$_2$O

Mechanism

*The nitronium ion is also called a **nitryl cation**.*

- The role of the concentrated sulfuric acid is to generate the **nitronium ion**, NO_2^+, as the active electrophile:

$$HNO_3 + H_2SO_4 \longrightarrow H_2NO_3^+ + HSO_4^-$$
$$H_2NO_3^+ \longrightarrow NO_2^+ + H_2O$$

- The powerful NO_2^+ electrophile then reacts with benzene.

*This is **electrophilic substitution**.*

attack of NO_2^+ electrophile → proton loss → product + H$^+$

- The H$^+$ formed regenerates a molecule of H$_2$SO$_4$.

$$H^+ + HSO_4^- \longrightarrow H_2SO_4$$

*The H$_2$SO$_4$ therefore acts as a **catalyst**.*

- The H$_2$SO$_4$ molecule reacts with more nitric acid to form more nitronium ions.

Chemistry of organic functional groups

Acylation of arenes

AQA M4

Acylation reactions of arenes are commonly known as *Friedel-Crafts* reactions. These are very important reactions industrially as they provide a means of introducing an acyl group, such as $CH_3C=O$, onto the benzene ring.

Acylation of benzene

A halogen carrier such as aluminium chloride, $AlCl_3$, is required to generate a reactive electrophile.

benzene + RCOCl (halogen carrier, e.g. $AlCl_3$, room temperature) → phenyl ketone + HCl

> R represents an alkyl group. By using different acyl chlorides, different acyl groups can be substituted onto the benzene ring.

Mechanism

Using an acyl chloride, in the presence of $AlCl_3$ as a halogen carrier, an **acylium ion** is generated.

$$RCOCl + AlCl_3 \longrightarrow RC^+=O + AlCl_4^-$$

- This acylium ion acts as a powerful electrophile which reacts with benzene.

$RC^+=O$ substitutes for H^+

Attack of $RC^+=O$ electrophile → intermediate → Proton loss → phenyl ketone + H^+

- The H^+ formed generates more $AlCl_3$.

$$H^+ + AlCl_4^- \longrightarrow AlCl_3 + HCl$$

- The $AlCl_3$ reacts with more of the acyl chloride RCOCl to generate more acylium ions. The **$AlCl_3$** therefore acts as a **catalyst**.

Progress check

1 What is the common type of reaction of arenes?

2 Benzene reacts with bromine, nitric acid and chloroethane.
 (a) Using C_6H_6 to represent benzene and C_6H_5 to represent the phenyl group, write balanced equations for each of these reactions.
 (b) State the essential conditions that are needed for each of these reactions.

1 Electrophilic substitution.
2 (a) $C_6H_6 + Br_2 \longrightarrow C_6H_5Br + HBr$
 $C_6H_6 + HNO_3 \longrightarrow C_6H_5NO_2 + H_2O$
 $C_6H_6 + C_2H_5Cl \longrightarrow C_6H_5C_2H_5 + HCl$
 (b) With bromine, a halogen carrier such as Fe.
 With nitric acid, concentrated sulfuric acid at 55°C.
 With chloroethane, a halogen carrier such as Fe.

Chemistry of organic functional groups

4.8 Amines

> **After studying this section you should be able to:**
> - explain the relative basicities of ethylamine and phenylamine
> - describe the reactions of primary amines with acids to form salts
> - describe the preparation of primary amines from haloalkanes and nitriles
> - describe the formation of phenylamine by reduction of nitrobenzene
> - describe the preparation by acylation of amines of amides

LEARNING SUMMARY

Aliphatic and aromatic amines

AQA   M4

Amines are organic compounds containing nitrogen derived from ammonia, NH_3. Amines are classified as primary, secondary or tertiary amines depending on how many of the hydrogen atoms in ammonia have been replaced by organic groups. The diagram below shows examples of aliphatic and aromatic amines.

Aliphatic amines

	1 hydrogen replaced	2 hydrogens replaced	3 hydrogens replaced
	primary amine	*secondary amine*	*tertiary amine*
NH_3	H_3C-NH_2	$H_3C\!\!-\!\!NH\!\!-\!\!CH_3$	$(H_3C)_3N$
ammonia	methylamine	dimethylamine	trimethylamine

Aromatic amine

C$_6$H$_5$—NH_2 phenylamine

Amines in nature

Amines are found commonly in nature. They are weak bases and the shortest chain amines, e.g. ethylamine $C_2H_5NH_2$, smell of fish. Some diamines, such as putrescine $H_2N(CH_2)_4NH_2$, and cadaverine $H_2N(CH_2)_5NH_2$, are found in decaying flesh.

Polarity

Notice the similarity with ammonia.

The presence of an electronegative nitrogen atom in amines results in polar molecules. The diagram shows the polarity of the primary amine methylamine.

$$H_3C-\overset{\delta-}{N}H_2 \quad H^{\delta+}$$

Physical properties of amines

As with other polar functional groups, the solubility of amines in water decreases with increasing carbon chain length as the non-polar contribution to the molecule becomes more important.

The amino group dominates the physical properties of short-chain amines.

Hydrogen bonding takes place between amine molecules, resulting in:
- higher melting and boiling points than alkanes of comparable relative molecular mass
- solubility in water – amines with fewer than six carbons mix with water in all proportions.

Chemistry of organic functional groups

Amines as bases

Amines are the organic bases.

Amines are weak bases because they only partially associate with protons:

$$RNH_2 + H^+ \rightleftharpoons RNH_3^+$$

The basic strength of an amine measures its ability to **accept** a proton, H^+. This depends upon:

- the size of the $\delta-$ charge on the amino nitrogen atom
- the availability of the nitrogen lone pair.

The basicity of aliphatic and aromatic amines

Aliphatic amines are stronger bases than ammonia; aromatic amines are substantially weaker.

Amines can also act as ligands with transition metal ions forming complex ions.

> **KEY POINT**
> - Electron-donating groups, e.g. alkyl groups, **increase** the basic strength.
> - Electron-withdrawing groups, e.g. C_6H_5, **decrease** the basic strength.

electron flow **away from** nitrogen ←
decrease in electron density of nitrogen lone pair — $C_6H_5-NH_2$

$H-NH_2$ (ammonia)

inductive effect causes electron flow **towards** nitrogen →
$H_3C \rightarrow NH_2$ — increase in electron density of nitrogen lone pair

basicity increases →

Organic ammonium salts

Amines react by the usual 'base reactions' producing organic ammonium salts.

Amines are **neutralised by acids** forming salts.

In the example below, methylamine is neutralised by hydrochloric acid forming a primary **ammonium salt**.

A proton, H^+, is added to the amino nitrogen atom.

$$CH_3NH_2 + HCl \longrightarrow CH_3NH_3^+Cl^-$$
methylammonium chloride

On evaporation of water, the primary ammonium salt crystallises out as an ionic compound.

Preparation of amines

Preparation of aromatic amines from nitroarenes

Aromatic amines can be prepared by **reducing** a nitroarene. For example, nitrobenzene is reduced by Sn in concentrated HCl to form phenylamine.

Other reducing agents can also be used:
- *H_2/Ni catalyst*
- *$LiAlH_4$ in dry ether*

$LiAlH_4$ is an almost universal reducing agent and works in most examples of organic reduction.

$C_6H_5-NO_2 + 6[H] \xrightarrow[\text{2. NaOH(aq)}]{\text{1. Sn / conc. HCl reflux}} C_6H_5-NH_2 + 2H_2O$

phenylamine

- Using tin and hydrochloric acid, the salt $C_6H_5NH_3^+Cl^-$ is formed.
- The amine is obtained from this salt by adding aqueous alkali.

Preparation of aliphatic amines from nitriles

Aliphatic amines can be prepared by **reducing** a nitrile with a suitable reducing agent (e.g. Na in ethanol; $LiAlH_4$ in dry ether).

$$CH_3CH_2C\equiv N + 4[H] \longrightarrow CH_3CH_2CH_2NH_2$$

Chemistry of organic functional groups

Reactions of amines with halogenoalkanes

AQA M4

Formation of a primary amine by nucleophilic substitution

Amines act as **nucleophiles** with organic compounds such as halogenoalkanes.

Halogenoalkanes react with **excess ammonia** in hot ethanol, forming a primary amine. The reaction with bromomethane is shown below:

$$CH_3Br + 2NH_3 \longrightarrow CH_3NH_2 + NH_4^+Br^-$$

The reaction takes place in two steps:

- Nucleophilic substitution with ammonia forms an alkylammonium ion:

$$CH_3Br + NH_3 \longrightarrow CH_3NH_3^+ + Br^-$$

- Proton transfer with ammonia forms a primary amine:

$$CH_3NH_3^+Br^- + NH_3 \rightleftharpoons CH_3NH_2 + NH_4^+Br^-$$

Key points from AS

- Nucleophilic substitution reactions of halogenoalkanes
  Revise AS pages 115–117

Mechanism and further substitution

The mechanism for the formation of a primary amine from bromoethane is shown below.

[Mechanism diagram showing nucleophilic substitution of H_3C-CH_2-Br by $:NH_3$ to form $H_3C-CH_2-N^+H_3 + Br^-$, followed by proton exchange with $:NH_3$ to give $H_3C-CH_2-NH_2 + NH_4^+Br^-$]

With an **excess** of **ammonia**, the reaction essentially stops at this stage.

With an **excess** of the **halogenoalkane**, further substitution takes place, which continues until a quaternary ammonium salt is obtained.

$$RNH_2 \xrightarrow{RX} R_2NH \xrightarrow{RX} R_3N \xrightarrow{RX} R_4N^+X^-$$
primary amine — secondary amine — tertiary amine — quaternary ammonium salt

Quaternary ammonium salts are used as cationic surfactants in fabric softeners.

Each reaction stage proceeds by **nucleophilic substitution**.

The mechanism for the final stage in the formation of the quaternary salt tetramethylammonium bromide is shown below.

[Mechanism showing $H_3C-C(CH_3)_2-Br$ reacting with $:N(CH_3)_3$ via nucleophilic substitution to give $[H_3C-N(CH_3)_3]^+ Br^-$]

- This mechanism stops at the quaternary salt as there are no further protons to be exchanged.

Chemistry of organic functional groups

Amides

AQA M4

Amides

Primary amides have the general formula $RCONH_2$ – R is an alkyl group.

The first six members of the **amides** homologous series are shown below.

structural formula	molecular formula	name
$HCONH_2$	CH_3NO	methanamide
CH_3CONH_2	C_2H_5NO	ethanamide
$CH_3CH_2CONH_2$	C_3H_7NO	propanamide
$CH_3CH_2CH_2CONH_2$	C_4H_9NO	butanamide
$CH_3CH_2CH_2CH_2CONH_2$	$C_5H_{11}NO$	pentanamide
$CH_3CH_2CH_2CH_2CH_2CONH_2$	$C_6H_{13}NO$	hexanamide

Types of amide

primary amide — secondary amide — tertiary amide

Preparation of amides

The amide group is stable and relatively unreactive. Preparation directly from a carboxylic acid is difficult (although this type of reaction is essentially what happens when peptides form from amino acids – see pages 118 and 124).

Amides are generally prepared by the reaction of ammonia and amines with **acyl chlorides** or with acid anhydrides – these are much more reactive than carboxylic acids (see pages 100–101).

- Reaction with ammonia produces a **primary amide**.

    $CH_3COCl + NH_3 \longrightarrow CH_3CONH_2 + HCl$

    (Then $NH_3 + HCl \longrightarrow NH_4Cl$)

- Reaction with a primary amine produces a **secondary amide**.

    $CH_3COCl + CH_3NH_2 \longrightarrow CH_3CONHCH_3 + HCl$

    (Then $NH_3 + HCl \longrightarrow NH_4Cl$)

> The acyl chloride is often prepared initially from a carboxylic acid. The acyl chloride can then be reacted with ammonia or amines to form amides.

An amide can be prepared from a carboxylic acid by first converting the carboxylic acid into an acyl chloride (see pages 100–101 for more details).

Progress check

1. Explain why ethylamine is a stronger base than phenylamine.
2. (a) Write equations for the conversion of:
    - (i) benzene to nitrobenzene
    - (ii) nitrobenzene to phenylamine
3. (a) Write down the formula of the organic product formed between:
    - (i) bromoethane and excess ammonia
    - (ii) ammonia and excess bromoethane.

1 The ethyl group has a positive inductive effect which reinforces the electron density of the nitrogen atom of the amine. This lone pair on the nitrogen will attract protons more strongly.
The phenyl group has a negative inductive effect which reduces the electron density on the nitrogen atom of the amine. This lone pair on the nitrogen will attract protons less strongly.
2 (a) (i) $C_6H_6 + HNO_3 \longrightarrow C_6H_5NO_2 + H_2O$ (ii) $C_6H_5NO_2 + 6[H] \longrightarrow C_6H_5NH_2 + 2H_2O$
3 (a) (i) $C_2H_5NH_2$ (ii) $(C_2H_5)_4N^+Br^-$

Chemistry of organic functional groups

Sample question and model answer

This question looks at the structure and reactions of benzene.

(a) The average enthalpy of hydrogenation of a single C–C bond is −120 kJ mol^{-1}.

(i) Assuming that benzene consists of a ring with three separate double bonds, predict the enthalpy change for the hydrogenation of benzene to form cyclohexane.

> *This is an easy mark. Don't forget to include the sign!*

3×-120 kJ mol^{-1} = -360 kJ mol^{-1} ✓

(ii) The actual enthalpy of hydrogenation of benzene is −205 kJ mol^{-1}. Using this information and your answer to (i), what conclusion can be drawn about the stability of the benzene ring? Use an enthalpy level diagram to illustrate your answer.

Benzene has extra stability arising from delocalisation of π-electrons ✓

> *An enthalpy diagram is a good way of showing the delocalisation energy of delocalised benzene.*
>
> *1 mark here is awarded for the relative positions of the delocalised and Kekulé structures of benzene.*

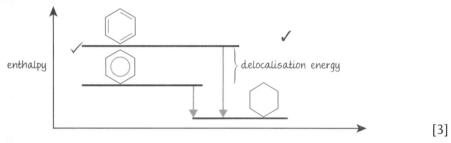

[3]

(b) Benzene can be nitrated to form nitrobenzene. Give the reagents, the equation and the conditions for this reaction.

> *This is standard bookwork that must be learnt. Other reactions could have been chosen (e.g. bromination).*
>
> *Remember that 'reagents' are the chemicals 'out of the bottle' that are reacted with the organic compound. You must give the full name or formula for any 'reagent'.*

Reagent(s): concentrated HNO$_3$/H$_2$SO$_4$ ✓

Equation: C$_6$H$_6$ + HNO$_3$ ⟶ C$_6$H$_5$NO$_2$ + H$_2$O ✓

Conditions: warm to 55°C ✓ [3]

(c) For the nitration of benzene:

(i) write the formula of the electrophile

NO$_2^+$ ✓

(ii) write the equation to show the formation of the electrophile

HNO$_3$ + H$_2$SO$_4$ ⟶ NO$_2^+$ + HSO$_4^-$ + H$_2$O ✓

> *More standard bookwork. An alternative response here would be:*
>
> HNO$_3$ + 2H$_2$SO$_4$ → NO$_2^+$ + 2HSO$_4^-$ + 2H$_2$O

(iii) state the type of mechanism

electrophilic substitution ✓

(iv) outline the mechanism.

The sulfuric acid catalyst is then regenerated:

H$^+$ + HSO$_4^-$ ⟶ H$_2$SO$_4$ ✓

> *Notice how precise you need to be to score all three marks:*
> - *arrow to electrophile ✓*
> - *correct intermediate ✓*
> - *arrow on C–H for loss of H$^+$ ✓*

[7]

[Total: 13]

Chemistry of organic functional groups

Practice examination questions

1 Compounds **A**, **B** and **C** are structural isomers, each containing a carbonyl group.

The compounds have a relative molecular mass of 72 and the following percentage composition by mass: C, 66.7%; H, 11.1%; O, 22.2%.

  (a) Calculate the molecular formula of the isomers. [2]

  (b) What are the structural formulae of **A**, **B** and **C**? [3]

  (c) Name a reagent that reacts with **A**, **B** and **C**. State the observation. [2]

  (d) Name a reagent that reacts with only two of the isomers **A**, **B** and **C**. Identify which isomers react and state the observation. [3]

  (e) One of the isomers **A**, **B** and **C** is reacted with $NaBH_4$ to form a product that has optical isomers. Identify this product and the isomer that has been reacted with $NaBH_4$. [2]

  [Total: 12]

2 (a) Compound **D**, $CH_3CH_2COOCH_2CH_3$, is an ester.

  (i) Name compound **D**.

  (ii) Compound **D** was hydrolysed by heating with aqueous sodium hydroxide. An alcohol **E** was formed, together with another organic product, **F**. Give the structural formulae of compounds **E** and **F**. [4]

  (b) Compound **G** is a di-ester formed by the reaction of ethane-1,2-diol and methanoic acid.

  (i) Draw the structural formula of compound **G**.

  (ii) Identify a substance that could catalyse this reaction. [2]

  (c) Compound **H** is a tri-ester. Complete and balance the equation below for the formation of the tri-ester **H** by esterification.

$$\longrightarrow \begin{array}{l} CH_3COO-CH_2 \\ CH_3COO-CH \\ CH_3COO-CH_2 \end{array}$$

  tri-ester **H** [3]

  [Total: 9]

3 An aromatic hydrocarbon has a relative molecular mass of 106 and has the following composition by mass: C, 90.56%; H, 9.44%.

  (a) (i) Deduce the molecular formula of the aromatic hydrocarbon.

  (ii) Draw structures for all the structural isomers of this aromatic hydrocarbon. [7]

  (b) In the presence of a sulfuric acid, one of these isomers, **I**, reacts with nitric acid to give only one mononitro product **J**.

  (i) Deduce which of the isomers in (a)(ii) is isomer **I**.

  (ii) Draw the structure of **J**. [2]

  [Total: 9]

Chemistry of organic functional groups

Practice examination questions (continued)

4 Methylbenzene can be converted into phenylamine using the reaction scheme below.

CH$_3$-C$_6$H$_5$ →(Step 1) CH$_3$-C$_6$H$_4$-NO$_2$ →(Step 2) CH$_3$-C$_6$H$_4$-NH$_2$

(a) For Step 1:
   (i) name the mechanism
   (ii) state the reagents required. [2]

(b) For the reduction in Step 2:
   (i) state the reagents required
   (ii) write a balanced equation (use [H] to represent the reducing agent). [3]

[Total: 5]

5 Compound **C**, CH$_3$CH$_2$CH$_2$COOCH(CH$_3$)$_2$, can be prepared using the reaction scheme below.

CH$_3$CH$_2$CH$_2$CHO →(Step 1) **A** + **B** ←(Step 2) CH$_3$COCH$_3$

Step 3 ↓

CH$_3$CH$_2$CH$_2$COOCH(CH$_3$)$_2$
C

(a) Show the structural formulae of compounds **A** and **B**. [2]

(b) For each of the three steps, give the reagents and conditions and name the type of reaction. [9]

[Total: 11]

Chapter 5
Polymers, analysis and synthesis

The following topics are covered in this chapter:

- Amino acids
- Polymers, proteins and nucleic acids
- Chromatography
- Analysis
- NMR spectroscopy
- Organic synthetic routes

5.1 Amino acids

After studying this section you should be able to:

- describe the acid–base properties of amino acids and the formation of zwitterions
- explain the formation of polypeptides and proteins as condensation polymers of amino acids
- describe the acid hydrolysis of proteins and peptides

Amino acids

AQA M4

A typical amino acid is an organic molecule with both acidic and basic properties comprising:

- a basic amino group, $-NH_2$
- an acidic carboxyl group, $-COOH$.

There are 22 naturally occurring amino acids.

Each amino acid has a unique organic R group or side chain. The formula of a typical amino acid is shown below.

$$H_2N-\underset{\underset{H}{|}}{\overset{\overset{R}{|}}{C}}-COOH$$

Examples of some amino acids are shown below:

$$H_2N-\underset{\underset{H}{|}}{\overset{\overset{H}{|}}{C}}-COOH \qquad H_2N-\underset{\underset{H}{|}}{\overset{\overset{CH_3}{|}}{C}}-COOH \qquad H_2N-\underset{\underset{H}{|}}{\overset{\overset{CH(CH_3)_2}{|}}{C}}-COOH$$

R = H   R = CH_3 R = $CH(CH_3)_2$
glycine   *alanine*   *valine*

Acid–base properties of amino acids

AQA M4

Isoelectric points and zwitterions

Each amino acid has a particular pH called the **isoelectric point** at which the overall charge on an amino acid molecule is zero.

Examples of isoelectric points

amino acid	aspartic acid	glycine	histidine	arginine
isoelectric point	3.0	6.1	7.6	10.8

At the isoelectric point, the amino acid exists in equilibrium with its zwitterion form.

At the isoelectric point, an amino acid exists as a zwitterion:

- the **carboxyl** group **has donated** a proton to the **amino** group, which form a positive NH_3^+ ion.

Polymers, analysis and synthesis

A **zwitterion** is a dipolar ion with both positive and negative charges in different parts of the molecule.

$$H_3\overset{+}{N}-\underset{H}{\overset{R}{C}}-C\overset{O}{\underset{O^-}{\diagup}}$$

zwitterion – two ions in one molecule

Amino acids as bases

In strongly **acidic** conditions a **positive ion** forms:
- an amino acid behaves as a **base**
- the COO^- ion gains a proton.

$$H_3\overset{+}{N}-\underset{H}{\overset{R}{C}}-C\overset{O}{\underset{O^-}{\diagup}} + H^+ \longrightarrow H_3\overset{+}{N}-\underset{H}{\overset{R}{C}}-C\overset{O}{\underset{OH}{\diagup}}$$

positive ion

Because of their reactions with strong acids and strong bases, amino acids act as buffers and help to stabilise the pH of living systems.

Amino acids as acids

In strongly **alkaline** conditions a **negative ion** forms:
- an amino acid behaves as an **acid**
- the NH_3^+ ion loses a proton.

$$H_3\overset{+}{N}-\underset{H}{\overset{R}{C}}-C\overset{O}{\underset{O^-}{\diagup}} + OH^- \longrightarrow H_2N-\underset{H}{\overset{R}{C}}-C\overset{O}{\underset{O^-}{\diagup}} + H_2O$$

negative ion

> **KEY POINT**
> - At the **isoelectric point**, the amino acid is **neutral**.
> - At a pH more **acidic** than the isoelectric point, the amino acid forms a **positive ion**.
> - At a pH more **alkaline** than the isoelectric point, the amino acid forms a **negative ion**.

Physical properties of amino acids

AQA M4

Solid amino acids

Solid amino acids have higher melting points than expected from their molecular masses and structure. This suggests that amino acids crystallise in a giant lattice with strong electrostatic forces between the **zwitterions**.

Optical isomers

With the exception of aminoethanoic acid (*glycine*), H_2NCH_2COOH, all amino acids have a chiral centre and are optically active.

Polymers, analysis and synthesis

Polypeptides and proteins

AQA M4

In nature, individual amino acids are linked together in chains as **polypeptides** and **proteins** (see page 124).

The diagram below shows the condensation of the amino acids glycine and alanine to form a **dipeptide**. The amino acids are bonded together by a **peptide link**.

> A protein is formed by condensation polymerisation of amino acids. See page 124.

$$H_2N-\underset{\underset{H}{|}}{\overset{\overset{H}{|}}{C}}-C\overset{\nearrow O}{\underset{\searrow OH}{}} + H_2N-\underset{\underset{H}{|}}{\overset{\overset{CH_3}{|}}{C}}-C\overset{\nearrow O}{\underset{\searrow OH}{}} \longrightarrow H_2N-\underset{\underset{H}{|}}{\overset{\overset{H}{|}}{C}}-\overset{\overset{O}{\|}}{C}-N-\underset{\underset{H}{|}}{\overset{\overset{CH_3}{|}}{C}}-C\overset{\nearrow O}{\underset{\searrow OH}{}} + H_2O$$

glycine　　　　　alanine　　　　　　　　　　peptide link

Each peptide link forms:

> A polypeptide is the name given to a short chain of amino acids linked by peptide bonds.
> A protein is simply the name given to a long-chain polypeptide.

- between the **carboxyl group** of glycine and the **amino group** of alanine
- with loss of a water molecule in a **condensation reaction**.

Further condensation reactions between amino acids build up a **polypeptide** or **protein**.

- For each amino acid added to a protein chain, one water molecule is lost.
- Most common proteins contain more than 100 amino acids.
- Each protein has a unique sequence of amino acids and a complex three-dimensional shape, held together by intermolecular bonds, including hydrogen bonds.

Hydrolysis of proteins

AQA M4

Hydrolysis breaks down a protein into its separate amino acids.

- The protein is refluxed with 6 mol dm^{-3} HCl(aq) for 24 hours.
- The resulting solution is neutralised.

The equation below shows the hydrolysis of a dipeptide.

> Proteins can also be hydrolysed with hot aqueous alkali.

peptide link　　　　　　　　　　　　　　　　glycine　　　alanine

- Hydrolysis of a protein is the reverse process to the condensation polymerisation that forms proteins.
- Biological systems use enzymes to catalyse this hydrolysis, which takes place at body temperature without the need for acid or alkali.

Identifying the amino acids in a protein

To determine the amino acids present in a protein, the protein is first boiled with 6 mol dm^{-3} hydrochloric acid. The amino acids formed are separated using paper chromatography and made visible by spraying the paper with ninhydrin.

Each amino acid moves a different distance on the chromatography paper, making for easy identification of the amino acids in the protein.

Polymers, analysis and synthesis

Progress check

1 The isoelectric point at which the zwitterion exists of serine (R = –CH$_2$OH) is 5.7. Draw the form of the molecule in aqueous solutions of pH 3.0, pH 5.7 and pH 10.0.

2 Draw the structure of the tripeptide with the sequence alanine–serine–aspartic acid (alanine: R = –CH$_3$; aspartic acid: R = –CH$_2$COOH).

Answers:

1. At pH 3.0: H$_3$N$^+$–CH(CH$_2$OH)–COOH; at pH 5.7 (zwitterion): H$_3$N$^+$–CH(CH$_2$OH)–COO$^-$; at pH 10.0: H$_2$N–CH(CH$_2$OH)–COO$^-$.

2. H$_2$N–CH(CH$_3$)–CO–NH–CH(CH$_2$OH)–CO–NH–CH(CH$_2$COOH)–COOH

Polymers, analysis and synthesis

5.2 Polymers, proteins and nucleic acids

After studying this section you should be able to:

- describe the characteristics of addition polymerisation
- describe the characteristics of condensation polymerisation in polypeptides, proteins, polyamides and polyesters
- discuss the disposal of polymers

LEARNING SUMMARY

During the study of alkenes in AS Chemistry, you learnt about addition polymers. For A2 Chemistry, addition polymerisation is reviewed and compared with condensation polymerisation.

Monomers and polymers

A **polymer** is a compound comprising very large molecules that are multiples of simpler chemical units called **monomers**.

Monomers are small molecules that can combine together to form a single large molecule, called a **polymer**.

> Two processes lead to formation of a polymer.
> - **Addition polymerisation** – monomers react together forming the polymer **only**. There are no by-products.
> - **Condensation polymerisation** – monomers react together forming the polymer **and** a simple compound, usually water.

KEY POINT

Addition polymerisation

AQA M4

In addition polymerisation:

- the monomer is an **unsaturated** molecule with a double **C=C** bond
- the double bond is **lost** as the **saturated** polymer forms.

Key points from AS

- Addition polymerisation of alkenes
  Revise AS pages 109–110

Many different addition polymers can be formed using different monomer units based upon alkenes.

Make sure that you can show a repeat unit of a polymer from the monomer (and a monomer from a repeat unit).

Poly(propenamide) and poly(ethenol) are highly water-absorbent polymers because of the ability of the polar –OH and –CONH$_2$ groups to hydrogen bond with water.

120

Polymers, analysis and synthesis

Properties of monomers and addition polymers

The monomers are volatile liquids or gases. Polymers are solids.
This difference can be explained in terms of van der Waals' forces.
- The van der Waals' forces acting between the large polymer molecules are much stronger than those acting between the much smaller monomer molecules.

Addition polymers of hydrocarbon monomers and of halogenated monomers are very stable and inert, e.g. poly(ethene), poly(propene) and poly(phenylethene) (polystyrene), poly(chloroethene) (PVC) and poly(tetrafluoroethene) (PTFE). These polymers are insoluble in solvents and are non-biodegradable.

Condensation polymerisation

In condensation polymerisation, the formation of a bond between monomer units also produces a small molecule such as H_2O or HCl.

Condensation polymers can be divided into natural polymers and man-made (synthetic) polymers.

- Natural polymers in living organisms include proteins, cellulose, rayon and DNA.
- Man-made polymers include synthetic fibres such as polyamides (e.g. *nylon*) and polyesters (e.g. *terylene*).

Polyamides

Nylon, proteins and polypeptides are all polyamides.

Proteins and polypeptides are natural condensation polymers. The link between the amino acid monomer units is usually described as a **peptide link** but chemically this is identical to an **amide** group. Hence polypeptides and proteins are **natural polyamides**.

Nylon-6,6 was the first man-made condensation polymer and was synthesised as an artificial alternative to natural protein fibres such as wool and silk.

Compare the use of two monomers in the production of synthetic nylon-6,6 with the natural condensation polymerisation using amino acids only.

The principle used was to mimic the natural polymerisation process above but, instead of using an amino acid monomer with two different functional groups, **two chemically different monomers are usually used**:

- a dicarboxylic acid **A**   HO–C(=O)–☐–C(=O)–OH   ☐ carbon chain
- a diamine **B** H_2N–◯–NH_2 ◯ or ring structure

Many different polyamides can be made using different carbon chains or rings which bridge the double functional group.

Each diamine molecule bonds to a dicarboxylic acid molecule with loss of water molecule.
- the two monomers **A** and **B** join alternately: –A–B–A–B–A–B–A–B–

–C(=O)–☐–C(=O)–N(H)–◯–N(H)–C(=O)–☐–C(=O)–N(H)–◯–N(H)–

Polyamides can also be made using a diacyl chloride instead of a dicarboxylic acid.
- The greater reactivity of an acyl chloride results in easier polymerisation.
- Hydrogen chloride is lost instead of water.

Polymers, analysis and synthesis

Formation of nylon-6,6 from its monomers

Nylon-6,6 gets its name from the number of carbon atoms in each monomer (diamine first).

The diamine has 6 carbon atoms.

The dicarboxylic acid has 6 carbon atoms.

Hence: nylon-6,6.

The diacyl chloride for preparing nylon-6,6 would be ClOC(CH$_2$)$_4$COCl.

hexane-1,6-dioic acid

$$\text{HO-C(=O)-(CH}_2\text{)}_4\text{-C(=O)-OH}$$

1,6-diaminohexane

$$\text{H}_2\text{N-(CH}_2\text{)}_6\text{-NH}_2$$

n molecules of a dicarboxylic acid

n molecules of a diamine

$$\text{HO-C(=O)-(CH}_2\text{)}_4\text{-C(=O)-OH} \quad \text{H-N(H)-(CH}_2\text{)}_6\text{-N(H)-H}$$

polymerisation ↓

amide link

$$\left[\text{-C(=O)-(CH}_2\text{)}_4\text{-C(=O)-N(H)-(CH}_2\text{)}_6\text{-N(H)-} \right]_n + n\,\text{H}_2\text{O}$$

repeat unit — nylon-6,6

One water molecule forms for each amide bond formed

Formation of Kevlar from its monomers

Other polyamides include Kevlar, one of the hardest materials known. Kevlar is used for bulletproof vests, belts for radial tyres, cables and reinforced panels in aircraft and boats.

$$\text{HO-C(=O)-C}_6\text{H}_4\text{-C(=O)-OH} \quad \text{H-N(H)-C}_6\text{H}_4\text{-N(H)-H}$$

polymerisation ↓

$$\left[\text{-C(=O)-C}_6\text{H}_4\text{-C(=O)-N(H)-C}_6\text{H}_4\text{-N(H)-} \right]_n + n\,\text{H}_2\text{O}$$

Kevlar

Polyesters

Learn the principle behind condensation polymerisation.

In exams, you may be required to predict structures from unfamiliar monomers.

However, the principle is the same.

Polyesters are polymers made by a condensation reaction between monomers with formation of an ester group as the linkage between the molecules.

The first man-made polyester produced was *Terylene*. As with polyamides, polyesters are used as fibres for clothing.

As with artificial polyamides, man-made polyesters are usually made from **two** different monomers:

- a dicarboxylic acid   HO-C(=O)-▭-C(=O)-OH   ▭ carbon chain or ● ring structure
- a diol   H-O-●-O-H

This is the same basic principle as for polyamides.

Each diol molecule bonds to a dicarboxylic acid molecule with loss of a water molecule.

Many different polyesters can be made by using different carbon chains or rings bridging the double functional group.

Polymers, analysis and synthesis

Formation of Terylene from its monomers

'Terylene' is a trade name for the polymer 'polyethylene terephthalate' (PET). PET is used to make the plastic for many drink bottles.

Disposal of polymers

AQA M4

Solid domestic and industrial waste contains a high percentage of polymers.

Disposal requires waste management strategies such as:

- incineration – to reduce waste bulk and to generate energy
- recycling – to preserve natural oil-produced finite resources produced from oil.

Problems with addition polymers (polyalkenes)

Addition polymers are non-polar and chemically inert.
This creates potential environmental problems during disposal of polymers.
Disposal by landfill causes long-term problems.

- Addition polymers are **non-biodegradable** and take many years to break down.

Disposal by burning can produce toxic fumes.

- Depolymerisation produces poisonous monomers.
- Disposal of poly(chloroethene) (PVC) by incineration can lead to the formation of very toxic dioxins if the temperature is too low.

Condensation polymers

Condensation polymers are polar.

- Condensation polymers are broken down naturally by acid and alkaline **hydrolysis** into their monomer units. They are therefore biodegradable and easier to dispose of than addition polymers.

123

Polymers, analysis and synthesis

Progress check

1 (a) Draw the structure of the monomer needed to make poly(tetrafluoroethene).
   (b) Draw the repeat unit of the polymer perspex, made from the monomer methyl 2-methylpropenoate, shown below.

   $$\begin{array}{c} H \\ \end{array} C=C \begin{array}{c} COOCH_3 \\ CH_3 \end{array}$$

2 (a) Show the structures of the two monomers needed to make nylon-4,6.
   (b) Draw a short section of nylon 4,6 and show its repeat unit.

3 Draw the structures of the two monomers needed to make the polyester shown below.

   $$\left[\begin{array}{c} O \\ \| \\ C \end{array}\!\!-\!\!\bigcirc\!\!-\!\!\begin{array}{c} O \\ \| \\ C \end{array}\!\!-\!\!O\!\!-\!\!\bigcirc\!\!-\!\!O \right]_n$$

Answers:

1 (a) F₂C=CF₂
  (b) $\left[\begin{array}{c} H \ \ CH_3 \\ -C-C- \\ H \ \ COOCH_3 \end{array}\right]_n$

2 (a) H₂N–(CH₂)₄–NH₂ and HOOC–(CH₂)₄–COOH
  (b) $\left[\begin{array}{c} H \ \ \ \ \ \ \ \ \ H \\ -N-(CH_2)_4-N-C-(CH_2)_4-C- \\ \| \ \ \ \ \ \ \ \ \ \ \ \ \ \ \ \| \\ O \ \ \ \ \ \ \ \ \ \ \ \ \ \ \ O \end{array}\right]_n$

3 HO–⌬–OH   HOOC–⌬–COOH

Proteins and polypeptides

AQA M4

Proteins and polypeptides

A protein is a chain of many amino acids linked together with 'peptide bonds' formed by **condensation polymerisation**.

Each amino acid molecule has both **amino** and **carboxyl** groups.
A **peptide** bond forms, with loss of a water molecule, between:

- the **amino** group of one amino acid molecule and
- the **carboxyl** group of another amino acid molecule.

The diagram below shows how amino acid molecules are linked together during condensation polymerisation.

$$H-N-\underset{\underset{H}{|}}{\overset{\overset{R}{|}}{C}}-\underset{}{\overset{\overset{O}{\|}}{C}}-OH \quad H-N-\underset{\underset{H}{|}}{\overset{\overset{R}{|}}{C}}-\underset{}{\overset{\overset{O}{\|}}{C}}-OH \quad H-N-\underset{\underset{H}{|}}{\overset{\overset{R}{|}}{C}}-\underset{}{\overset{\overset{O}{\|}}{C}}-OH \quad H-N-\underset{\underset{H}{|}}{\overset{\overset{R}{|}}{C}}-\underset{}{\overset{\overset{O}{\|}}{C}}-OH$$

↓

peptide link → , repeat unit

$$-N-\overset{R}{\underset{H}{C}}-\overset{O}{\underset{}{C}}-N-\overset{R}{\underset{H}{C}}-\overset{O}{\underset{}{C}}-N-\overset{R}{\underset{H}{C}}-\overset{O}{\underset{}{C}}\left[-N-\overset{R}{\underset{H}{C}}-\overset{O}{\underset{}{C}}-\right]$$

H₂O   H₂O   H₂O

> Although a 'repeat unit' is shown on the diagram, the R group may be different in each unit. There are over 20 different amino acids, each having a different R group.

- Most common proteins contain more than 100 amino acids.
- Each protein has a unique sequence of amino acids and a complex three-dimensional shape, held together by intermolecular bonds including hydrogen bonds.

5.3 Chromatography

After studying this section you should be able to:

- understand what is meant by chromatography
- know that column chromatography separates by relative solubility and retention
- know that GLC separates mixtures of volatile liquids

Types of chromatography

AQA M4

Chromatography is an analytical technique that separates components in a mixture between a **mobile phase** and a **stationary phase**. The technique is especially useful for the purification of an organic substance.

- The **mobile phase** moves in a definite direction and may be a liquid (as in column chromatography) or a gas (as in gas-liquid chromatography, GLC).
- The **stationary phase** is the substance that is fixed in place during the chromatography. It may be a solid (as in column chromatography) or either a liquid or solid on a solid support (as in GLC).

Column chromatography

Column chromatography uses a stationary phase within a column. The liquid mobile phase is passed through the column and components are separated depending on their solubility in the mobile phase and retention in the stationary phase. Liquid chromatography can be run under pressure using very small packing particles. This is referred to as **high pressure liquid chromatography (HPLC)**.

Gas-Liquid Chromatography GLC

Gas-Liquid chromatography (GLC) is a separation technique in which the mobile phase is a gas. Gas chromatography is always carried out in a column. The column contains the stationary phase, which is either a high-boiling liquid or a solid on a solid, porous support. GLC can be used to separate the components in a mixture of volatile liquids

- The sample is injected into the column, which is heated to vaporise the components in the mixture.
- The gas mobile phase flushes the mixture along the column.
- As the mixture moves through the column, components are slowed down as they interact with the stationary phase. Different components are slowed down by different amounts, which causes them to separate.
- Each component leaves the column at a different time.

Polymers, analysis and synthesis

5.4 Analysis

After studying this section you should be able to:

- review from AS Chemistry the use of infra-red spectroscopy and mass spectrometry in organic analysis

LEARNING SUMMARY

Using infra-red (IR) spectroscopy in analysis

AQA M2, M4

Basic principles

Bonds in molecules naturally vibrate. Some bonds in molecules increase their vibrations by absorbing energy from IR radiation. Different bonds absorb different frequencies of IR radiation.

The frequency of IR absorption is measured in wavenumbers, units: cm⁻¹.

An IR spectrum is obtained by passing a range of IR frequencies through a compound. As energy is taken in, **absorption peaks** are produced. The frequencies of the absorption peaks can be matched to those of known bonds to identify structural features in an unknown compound.

IR radiation has less energy than visible light.

> **KEY POINT**
> IR spectroscopy is useful for identifying the functional groups in a molecule.

Important IR absorptions

You don't need to learn the absorption frequencies – the data is provided.

bond	wavenumber/cm⁻¹
N–H (amines)	3300 – 3500
O–H (alcohols)	3230 – 3550
C–H	2850 – 3300
O–H (acids)	2500 – 3000
C≡N	2220 – 2260
C=O	1680 – 1750
C=C	1620 – 1680
C–O	1000 – 1300
C–C	750 – 1100

IR spectroscopy is used in some modern breathalysers for measuring the concentration of blood alcohol. A particular IR absorption identifies the presence of ethanol in the breath and the intensity of the peak is directly related to the ethanol level.

An infra-red spectrum is particularly useful for identifying:

- an **alcohol** from absorption of the O–H bond
- a **carbonyl** compound from absorption of the C=O bond
- a **carboxylic acid** from absorption of the C=O bond **and** broad absorption of the O–H bond.

Interpreting infra-red spectra

AQA M2, M4

Carbonyl compounds (aldehydes and ketones)
Butanone, CH₃COCH₂CH₃

- C=O absorption 1680 to 1750 cm⁻¹

126

Polymers, analysis and synthesis

IR spectroscopy is most useful for identifying C=O and O–H bonds.
Look for the distinctive patterns.

Alcohols
Ethanol, C_2H_5OH
- O–H absorption 3230 to 3500 cm^{-1}
- C–O absorption 1000 to 1300 cm^{-1}

O–H absorption at 3300 cm^{-1}
C–O absorption at 1010 cm^{-1}

Note that all these molecules contain C–H bonds, which absorb in the range 2840 to 3045 cm^{-1}.

Carboxylic acids
Propanoic acid, C_2H_5COOH
- Very broad O–H absorption 2500 to 3500 cm^{-1}
- C=O absorption 1680 to 1750 cm^{-1}

Very broad O–H absorption between 2500 and 3500 cm^{-1}
C=O absorption at 1680 cm^{-1}
C–O absorptions at 1200 and 1030 cm^{-1}

There are other organic groups (e.g. N–H, C=C) that absorb IR radiation but the principle of linking the group to the absorption wavenumber is the same.

Ester
Ethyl ethanoate, $CH_3COOC_2H_5$
- C=O absorption 1680 to 1750 cm^{-1}
- C–O absorption 1000 to 1300 cm^{-1}

C=O absorption at 1740 cm^{-1}
C–O absorptions at 1220 and 1020 cm^{-1}

The fingerprint region is unique for a particular compound.

Fingerprint region
- Between 1000 and 1550 cm^{-1}

Many spectra show a complex pattern of absorption in this range.
- This pattern can allow the compound to be identified by comparing its spectrum with spectra of known compounds.

Mass spectrometry

AQA M2, M4

Mass spectrometry can be used to determine relative atomic masses from a mass spectrum. Mass spectrometry is also used to determine relative molecular masses and to identify the molecular structures of organic compounds.

Molecular ions

The molecular ion peak is usually given the symbol M.

Organic molecules can be analysed using mass spectrometry.

In the mass spectrometer, organic molecules are bombarded with electrons. This can lead to the formation a **molecular ion**.

The mass spectrum also contains fragments of the molecule ion (see detail below).

The equation below shows the formation of a molecular ion from butanone, $CH_3COCH_2CH_3$.

$$H_3C-\overset{O}{\underset{\|}{C}}-CH_2CH_3 + e^- \longrightarrow \left[H_3C-\overset{O}{\underset{\|}{C}}-CH_2CH_3\right]^+ + 2e^-$$

molecular ion, *m/z*: 72

- The molecular ion peak, M, provides the relative molecular mass of the compound.

Fragment ions

In the conditions within the mass spectrometer, some molecular ions are fragmented by bond fission.

- The bond fission that takes place is a fairly random process:
  - different bonds are broken
  - a **mixture of fragment ions** is obtained.
- The mass spectrum contains both the molecular ion and the mixture of fragment ions.

Mass spectrometry of organic compounds is useful for identifying:

- the relative molecular mass
- parts of the skeletal formula.

The fragmentation pattern provides clues about the molecular structure of the compound.

Fragmentation of butane, C_4H_{10}

Bond fission forms a fragment ion and a radical. The diagram below shows that fission of the same C–C bond in a butane molecule can form two different fragment ions.

The mass spectrum of butane would contain peaks for these four ions:
- $C_4H_{10}^+$: $m/z = 58$
- $C_3H_7^+$: $m/z = 43$
- $C_2H_5^+$: $m/z = 29$
- CH_3^+: $m/z = 15$

$$H_3C-CH_2-CH_2\vdots CH_3^+ \quad m/z = 58$$

loss of 15 mass units → $H_3C-CH_2-CH_2^+$ + $^\bullet CH_3$ $m/z = 43$

loss of 43 mass units → $H_3C-CH_2-\overset{\bullet}{C}H_2$ + CH_3^+ $m/z = 15$

- Fragmentation of the molecular ion produces a fragment ion by loss of a radical.
- The mass spectrometer only detects the fragment ion.
- The fragment ion may itself fragment into another ion and radical.

Note that only ions are detected in the mass spectrum.

Uncharged species such as the radical cannot be deflected within the mass spectrometer.

The mass spectrum of butanone

The mass spectrum of butanone, $CH_3CH_2COCH_3$, is shown below. The m/z values of the main peaks have been labelled.

- base peak at m/z: 43
- m/z: 29
- m/z: 15
- m/z: 57
- M peak at m/z: 72
- M+1 peak at 73

The M+1 peak is a small peak 1 unit higher than the molecular ion peak.

The origin of the M+1 peak is the small proportion of carbon-13 in the carbon atoms of organic molecules.

The table below shows the identities of the main peaks.

m/z	ion	fragment lost
72	$CH_3COCH_2CH_3^+$	–
57	$CH_3CH_2CO^+$	$CH_3^\bullet$
43	CH_3CO^+	$CH_3CH_2^\bullet$
29	$CH_3CH_2^+$	$CH_3CO^\bullet$
15	CH_3^+	$CH_3CH_2CO^\bullet$

A common mistake in exams is to show both the fragmentation products as ions.

The most abundant peak in the mass spectrum is the base peak.

- The base peak in the mass spectrum of butanone has m/z: 43, formed by the loss of a $CH_3CH_2^\bullet$ radical:

$$\left[H_3C-\overset{O}{\underset{\|}{C}}-CH_2CH_3\right]^+ \longrightarrow \left[H_3C-\overset{O}{\underset{\|}{C}}\right]^+ + {}^\bullet CH_2CH_3$$

molecular ion   fragment ion   radical
m/z: 72 m/z: 43   M – 29

The acylium (RCO^+) ion is particularly stable and is often present as a high dominant peak.

The base peak is given a relative abundance of 100. Other peaks are compared with this base peak.

Polymers, analysis and synthesis

Common patterns in mass spectra

Different fragmentations are possible depending on the structure of the molecular ion.

- The table on p128 shows common fragment ions and fragmented radicals.
- The skill in interpreting a mass spectrum is in searching for known patterns, evaluating all the evidence and reassembling all the information into a molecular structure.
- You should also look for a peak at m/z 77 or loss of 77 units. This is a giveaway for the presence of a phenyl group, C_6H_5, in a molecule.

5.5 NMR spectroscopy

After studying this section you should be able to:

- understand that NMR spectroscopy is carried out with ^1H and ^{13}C
- predict the different types of proton and carbon present in a molecule from chemical shift values
- predict the relative numbers of each type of proton present from an integration trace
- predict the number of protons adjacent to a given proton from the spin-spin coupling pattern
- predict possible structures for a molecule from ^1H and ^{13}C spectra
- predict the chemical shifts and splitting patterns of the protons in a given molecule

Nuclear magnetic resonance

Nuclear Magnetic Resonance (NMR) spectroscopy is an extremely important modern method of analysis. It is used extensively in the pharmaceutical industry and in universities for identification and purity checking of organic compounds.

Key principles

The nucleus of an atom of hydrogen (i.e. a proton) has a magnetic spin. When placed in a strong electromagnetic field:

Only nuclei with an odd number of nucleons (neutrons + protons) possess a magnetic spin: e.g. ^1H, ^{13}C. Proton NMR spectroscopy is the most useful general purpose technique.

- the nucleus can absorb energy from the low energy **radio-frequency** region of the spectrum to move to a higher energy state
- **nuclear magnetic resonance** occurs as protons resonate between their spin energy states.

The different nuclear spin states in an applied magnetic field

The energy gap is equal to that provided by radio waves.

An NMR spectrum shows **absorption peaks** corresponding to the radio-frequency absorbed.

Chemical shift, δ

Electrons around the nucleus **shield** the nucleus from the applied field.

- The magnetic field at the nucleus of a particular proton is different from the applied magnetic field.

NMR spectroscopy is the same technology as used in MRI (magnetic resonance imaging) in medical body scanners. It is a non-invasive technique.

Protons in different environments absorb at different chemical shifts.

- Different radio-frequencies are absorbed, depending on the **environment** of the proton.
- **Chemical shift** is a measure of the magnetic field experienced by protons in different environments resulting from nuclear shielding.

Polymers, analysis and synthesis

> **KEY POINT**
> Chemical shift, δ, is measured relative to a standard: tetramethylsilane (TMS), $Si(CH_3)_4$.
> - The chemical shift of TMS is defined as δ = 0 ppm.

Carbon-13 NMR spectroscopy

Carbon-13 NMR spectroscopy allows the identification of carbon atoms in an organic molecule and is an important tool in the determination of structure in organic chemistry.

Typical chemical shifts

Chemical shifts indicate the chemical environment of the **carbon atoms** present. The table below shows typical chemical shift values for different types of carbon atom.

You don't need to learn these chemical shifts – the data is provided on exam papers.

type of carbon		chemical shift, δ/ppm
—C—C—		5–40
R—C—Cl or Br		10–70
R—C—C— ‖ O		20–50
R—C—N⟨		25–60
—C—O—	alcohol, ethers or esters	50–90
C=C		90–150
R—C≡N		110–125
(benzene ring)		110–160
R—C— ‖ O	esters or acids	160–185
R—C— ‖ O	aldehydes or ketones	190–220

The presence of an electronegative atom or group causes chemical shift 'downfield'. This is called 'deshielding'.

Notice the chemical shift caused by the carbonyl group, oxygen, halogens and a benzene ring.

- The actual chemical shift may be slightly different depending upon the actual environment of the carbon.

In a carbon–13 NMR spectrum, the size of the peak tells us nothing about the number of carbon atoms responsible. This is in contrast to proton NMR spectroscopy. However, carbon-13 spectrum is far more sensitive with a much larger range of chemical shift values. Unless there are carbon atoms present in an identical environment, each carbon atom is likely to show as a separate single signal.

131

Polymers, analysis and synthesis

Interpreting a carbon-13 NMR spectrum

A carbon-13 NMR spectrum of ethyl ethanoate, $CH_3COOCH_2CH_3$ shows absorptions at **four** chemical shifts showing carbon atoms in **four different environments**:

- The four chemical shifts can be matched to the table of chemical shifts on page 131 so that the carbon atom responsible for each absorption can be identified.

chemical shift/ppm	environment
171	$H_3C-C(=O)-O-CH_2-CH_3$ (C of C=O highlighted)
60	$H_3C-C(=O)-O-CH_2-CH_3$ (CH_2 highlighted)
21	$H_3C-C(=O)-O-CH_2-CH_3$ (H_3C- highlighted)
14	$H_3C-C(=O)-O-CH_2-CH_3$ (terminal CH_3 highlighted)

> **KEY POINT**
>
> Carbon-13 NMR spectroscopy is useful for identifying the **type** of environment of each carbon – the chemical shift.

Progress check

1. For each structure, predict the chemical shift of each carbon atom.
   (a) CH_3CH_2OH; (b) CH_3CH_2CHO; (c) CH_3COCH_3.

1 (a) CH_3CH_2OH, δ = 5–55 ppm; CH_3CH_2OH, δ = 50–70 ppm
(b) CH_3CH_2CHO, δ = 5–55 ppm; CH_3CH_2CHO, δ = 5–55 ppm; CH_3CH_2CHO, δ = 190–220 ppm
(c) CH_3COCH_3, δ = 5–55 ppm; CH_3COCH_3, δ = 190–220 ppm

Polymers, analysis and synthesis

Proton NMR spectroscopy

AQA M4

Proton NMR spectroscopy allows the identification of hydrogen atoms in an organic molecule and is an important tool in the determination of structure in organic chemistry.

Typical chemical shifts

Chemical shifts indicate the **types of protons** and **functional groups** present. The table below shows typical chemical shift values for protons in different chemical environments.

> The presence of an electronegative atom or group causes chemical shift 'downfield'. This is called 'deshielding'.
>
> Notice the chemical shift caused by the carbonyl group, oxygen, halogens and a benzene ring.

type of proton	δ/ppm
ROH	0.5–5.0
RCH_3	0.7–1.2
RNH_2	1.0–4.5
R_2CH_2	1.2–1.4
R_3CH	1.4–1.6
R—C(=O)—C—H	2.1–2.6
R—O—C—H	3.1–3.9
RCH_2Cl or Br	3.1–4.2
R—C(=O)—O—C—H	3.7–4.1
R\C=C/H (alkene)	4.5–6.0
R—C(=O)—H	9.0–10.0
R—C(=O)—O—H	10.0–12.0

> You don't need to learn these chemical shifts – the data is provided on exam papers.

- The actual chemical shift may be slightly different depending upon the actual environment of the proton.
- The chemical shift for O–H can vary considerably and depends upon concentration, solvent and other factors.

Polymers, analysis and synthesis

Interpreting low resolution NMR spectra

AQA M4

Low resolution NMR spectrum of ethanol

A low resolution NMR spectrum of ethanol shows absorptions at **three** chemical shifts, showing the **three different types of proton**:

> The relative areas of each peak are usually measured by running a second NMR spectrum as an *integration trace*.

> An NMR spectrum is obtained in solution. The solvent must be proton-free, usually CCl_4 or $CDCl_3$ is used.

- The **area** under each peak is in direct proportion to the **number of protons** responsible for the absorption.
- The three chemical shifts can be matched to the table of chemical shifts on page 133 so that the protons responsible for each absorption can be identified.

chemical shift/ppm	no. of protons	environment	type of proton
$\delta = 1.0$	3H	CH_3CH_2OH	CH_3 adjacent to a carbon chain
$\delta = 3.5$	2H	CH_3CH_2OH	CH_2 adjacent to $-O$
$\delta = 4.9$	1H	CH_3CH_2OH	OH

> **KEY POINT**
>
> A low resolution NMR spectrum is useful for identifying:
> - the **number** of different types of proton from the number of peaks
> - the **type** of environment of each proton from the chemical shift
> - **how many** protons of each type from the integration trace.

Interpreting high resolution NMR spectra

AQA M4

A high resolution NMR spectrum shows splitting of peaks into a pattern of sub-peaks.

Spin-spin coupling patterns:
- arise from interactions between protons on **adjacent** carbon atoms which have **different chemical shifts**
- indicate the number of **adjacent** protons.

> Different types of proton have different chemical shifts.

Equivalent protons (i.e. protons with the same chemical shift) will **not** couple with one another.

The **spin-spin coupling** pattern shows as a **multiplet** – a doublet, triplet, quartet, etc.

> A singlet is next to C.
> A doublet is next to CH.
> A triplet is next to CH_2.
> A quadruplet is next to CH_3.

> **KEY POINT**
>
> We can interpret the spin-spin coupling pattern using the **n+1 rule**.
> For **n adjacent protons**, the number of peaks in a multiplet = n+1.

The high resolution NMR spectrum of ethanol

The high resolution NMR spectrum of ethanol shows spin-spin coupling patterns – some of the signals have been split into multiplets:

Polymers, analysis and synthesis

The multiplicity (doublet, triplet, quartet) does not indicate the number of protons on that carbon. The number of protons is given by the integration trace.

Notice the different sizes of each sub-peak within the splitting pattern:
- doublet 1:1
- triplet 1:2:1
- quartet 1:3:3:1.

- The **chemical shift** identifies the **type of protons** responsible for each peak.
- The n+1 rule can be used to identify the **number of adjacent protons**.

chemical shift /ppm	environment	number of adjacent protons (n)	splitting pattern (n+1)
δ = 1.2	C**H**$_3$CH$_2$OH	2H	2+1 = 3: triplet
δ = 3.6	CH$_3$C**H**$_2$OH	3H	3+1 = 4: quartet
δ = 4.5	CH$_3$CH$_2$O**H**	–	singlet

- Note that O–H splitting is not normally seen owing to H-bonding or exchange with the solvent used.

Equivalent protons do not interact with each other.

- The three equivalent CH$_3$ protons in ethanol cause splitting of the adjacent CH$_2$ protons, but not amongst themselves.

> **KEY POINT**
> A high resolution NMR spectrum is useful for identifying the **number** of **adjacent** protons from the spin-spin coupling pattern.

Solvents

Deuterated solvents such as CDCl$_3$ are used in NMR spectroscopy. Any protons in a solvent such as CHCl$_3$ would produce a large proton absorption peak. By using CDCl$_3$, this absorption is absent and the spectrum is that of the organic compound alone.

Progress check

1. For each structure, predict the number of peaks in its low resolution NMR spectrum corresponding to the different types of proton and also the number of each different type of proton
   (e.g. CH$_3$CH$_2$OH has 3 peaks in the ratio 3:2:1).
   (a) CH$_3$OH (b) CH$_3$CH$_2$CHO (c) CH$_3$COCH$_3$ (d) (CH$_3$)$_2$CHOH.

2. The NMR spectrum of compound **X** (C$_2$H$_4$O) has a doublet at δ 2.1 and a quartet at δ 9.8.
   (a) Identify compound **X**. (Use the table of chemical shifts on page 133).
   (b) How many protons are responsible for each multiplet?

1 (a) 2 peaks in the ratio 3:1 (b) 3 peaks in the ratio 3:2:1 (c) 1 peak (d) 3 peaks in the ratio 6:1:1.
2 (a) ethanal, CH$_3$CHO. (b) doublet at δ 2.1, 3H; quartet at δ 9.8, 1H.

5.6 Organic synthetic routes

There are many variations possible and the schemes below could not include all reactions without appearing more complicated.

Organic chemists are frequently required to synthesise an organic compound in a multistage process. This is fundamental to the design of new organic compounds such as those needed for modern drugs to combat disease, a new dye or a new fibre. With so many reactions to consider, it is essential to see how different functional groups can be interconverted and summary charts are useful for showing these links. The flow-charts on the next pages show reactions drawn from both the AS and A2 parts of A Level Chemistry.

Construct a scheme of your own using only those reactions in your course.

In revision it is useful to draw your own schemes from memory.

Aliphatic synthetic routes

The scheme below is based upon two main sets of reactions:
- a set based around bromoalkanes
- a set based around carboxylic acids and their derivatives.

The two sets of reactions are linked via the oxidation of primary alcohols.
- A reaction scheme based upon a secondary alcohol would result in oxidation to a ketone only.

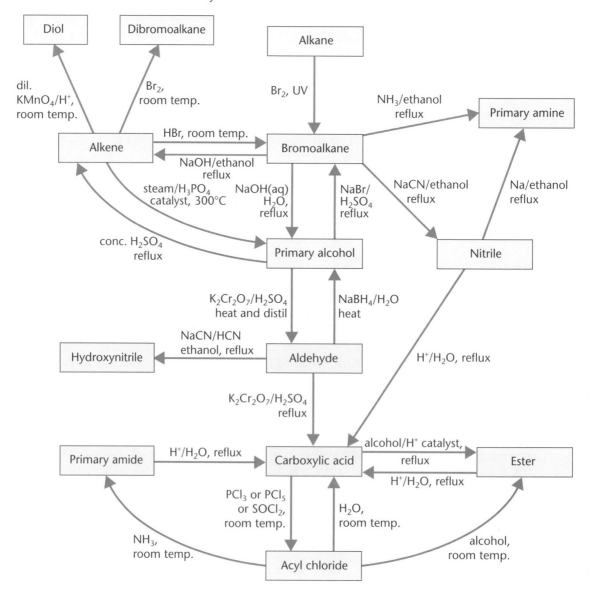

Polymers, analysis and synthesis

Aromatic synthetic routes

AQA M4

Compared with aliphatic organic chemistry at A Level, there are comparatively few aromatic reactions.

Benzene

- Reactions involving the benzene ring are mainly electrophilic substitution.

Benzene reactions:
- **alkylation**: RCl / AlCl$_3$, warm → gives R-substituted benzene
- **acylation**: RCOCl / AlCl$_3$, warm → gives phenyl ketone (C(=O)R)
- **nitration**: H$_2$SO$_4$ / HNO$_3$ conc., 55°C → gives nitrobenzene (NO$_2$)
- **reduction** of nitrobenzene: 1. Sn / conc HCl reflux 2. NaOH(aq) → gives aniline (NH$_2$)

It is essential, if you are to answer such questions, that you thoroughly learn suitable reagents and conditions for all the reactions in the AQA syllabus.

- Questions are often set in exams asking for reagents, conditions or products.
- More searching problems may expect a synthetic route from a starting material to a final product. The synthetic route may include several stages.

Polymers, analysis and synthesis

Sample question and model answer

The structures of the four esters with the molecular formula $C_4H_8O_2$ are shown below.

$$CH_3CH_2COOCH_3 \quad CH_3COOCH_2CH_3 \quad HCOOCH_2CH_2CH_3 \quad HCOOCH(CH_3)_2$$
$$\quad\quad A \quad\quad\quad\quad\quad\quad B \quad\quad\quad\quad\quad\quad\quad C \quad\quad\quad\quad\quad\quad\quad D$$

(a) For each structure predict the number of peaks in its NMR spectrum and the ratio of the number of protons responsible for each peak.

When we are predicting the number of peaks, we are identifying the number of different types of proton.

In compound D, $HCOOCH(CH_3)_2$, both the methyl groups (6H) are equivalent and will have the same chemical shift.

Ester A, $CH_3CH_2COOCH_3$, has 3 peaks ✓ in the ratio 3:2:3 ✓
Ester B, $CH_3COOCH_2CH_3$, has 3 peaks ✓ in the ratio 3:2:3 ✓
Ester C, $HCOOCH_2CH_2CH_3$, has 4 peaks ✓ in the ratio 1:2:2:3 ✓
Ester D, $HCOOCH(CH_3)_2$, has 3 peaks ✓ in the ratio 1:1:6 ✓ [8]

(b) The NMR spectrum of one of the four esters is shown below. Integration data is shown by each peak. Analyse the spectrum to find out which ester has produced it. Include all the supporting evidence from the spectrum.

Notice how the answer is methodical, going through each piece of evidence separately.

The combination of a triplet and quartet is a giveaway for a CH_3CH_2 combination.

The integration data supports either structure A or B:
  There are three types of proton in the ratio 3:2:3 ✓
The splitting pattern supports either structure A or B:
  The CH_3 group at δ = 1.14 ppm has been split into a triplet ✓ by 2 protons on an adjacent carbon atom. ✓
  The CH_2 group at δ = 2.29 ppm has been split into a quartet ✓ by 3 protons on an adjacent carbon atom. ✓
  The CH_3 group at δ = 3.68 ppm is a singlet ✓ so there can be no protons on an adjacent carbon. ✓

The integration data and splitting patterns point to either structure A or B.
The chemical shifts are conclusive though.

From chemical shifts,
  the CH_3 group at δ = 3.68 ppm must be adjacent to an O atom ✓
  the CH_2 group at δ = 2.29 ppm must be adjacent to a C=O group. ✓
Taking all the evidence together:
  the ester must be A, $CH_3CH_2COOCH_3$. ✓ [10]

(c) The IR and mass spectra can also be used to help confirm structures. What key IR absorptions and m/z values would you expect to see in the mass and IR spectra of ester A?

There are 4 marks here. The first is for the key information: the C=O absorption in the IR and the molecular ion peak in the mass spectrum. The second mark is for identification of another key feature.

IR: A C=O absorption at about 1700 cm^{-1}. ✓ A C–O absorption at 1000–1300 cm^{-1}. There should be no absorption above 3000 cm^{-1} as there is no –OH group present. ✓

Other fragment ions could have been chosen.

Mass spectrum: A molecular ion peak at m/z = 88 confirming the molecular mass. ✓ Fragments ions at m/z = 57 from $CH_3CH_2CO^+$; also at m/z = 31 for CH_3O^+. ✓ [4]

[Total: 22]

Polymers, analysis and synthesis

Practice examination questions

1 Valine, (CH$_3$)$_2$CH(NH$_2$)COOH, is an amino acid.

(a) What is the 'R' group in valine? [1]

(b) Valine reacts with an amino acid, **A**, to form the dipeptide below.

$$\begin{array}{c}\text{H}\\\text{H}\diagdown\;\;\;|\;\text{H}\quad\quad\;\;\;\text{O}\\\text{N—CH—C—N—CH—C}\diagup\\\text{H}\diagup\;\;|\;\;\;\;\|\;\;\;\;\;\;\;\;|\;\;\;\;\;\;\;\;\diagdown\text{OH}\\\text{CH}_3\;\;\text{O}\;\;\;\;\;\;\text{CH}\\\quad\quad\quad\text{H}_3\text{C}\diagup\;\;\diagdown\text{CH}_3\end{array}$$

(i) Draw a circle around the peptide linkage.

(ii) Draw the structure of the amino acid, **A**.

(iii) Valine can react with the amino acid, **A**, to form a different dipeptide from that shown above. Draw the structural formula of this other dipeptide. [3]

(c) Valine has a chiral centre and can exist as two optical isomers.

(i) State what is meant by a *chiral centre* and explain how a chiral centre gives rise to optical isomerism.

(ii) Draw diagrams to show the two optical isomers of valine. State the bond angle around the chiral centre. [5]

(d) In aqueous solution, valine exists as different ions at different pH values. The zwitterion exists in the pH range 3–10. Draw the displayed formula of the ion of valine present at pH values of 2.0, 7.0 and 12.0. [3]

(e) Compound **B** is an amino acid with the molecular formula C$_3$H$_7$NO$_3$. Draw the structure of amino acid **B**. [1]

[Total: 13]

2 Lactic acid has the structural formula CH$_3$CH(OH)COOH.

(a) What is the systematic name for lactic acid? [1]

(b) Lactic acid has optical isomers.

(i) What structural feature in lactic acid results in optical isomerism?

(ii) How does optical isomerism arise?

(iii) Draw a three-dimensional diagram to show the optical isomers of lactic acid. [4]

[Total: 5]

Polymers, analysis and synthesis

Practice examination questions (continued)

3 The repeat units of two polymers **C** and **D** are shown below.

(a) (i) Draw the structure of the monomer of polymer **C**.

(ii) Name the type of polymerisation. [2]

(b) (i) Draw the structure of the two monomers of polymer **D**.

(ii) Name the type of polymerisation. [3]

(c) Suggest why polymer **D** would be more likely to be hydrolysed than polymer **C**. [2]

[Total: 7]

4 Compounds **E** and **F** are both diols with the molecular formula $C_4H_{10}O_2$.

(a) The proton NMR spectrum of a diol **E**, $C_4H_{10}O_2$, is shown below. The numbers by each peak represent the integration trace. The peak at $\delta = 3.65$ ppm is due to hydroxyl protons, –OH.

(i) How many different types of proton are present in the diol?

(ii) The peaks at $\delta = 1.63$ ppm and $\delta = 0.90$ ppm result from a single alkyl group. Identify this group and explain the splitting pattern.

(iii) What can be deduced from the single peak at $\delta = 1.32$ ppm?

(iv) What can be concluded about the peak at $\delta = 3.65$ ppm?

(v) Show the structure of the diol **E**. [9]

(b) The proton NMR spectrum diol **F** is shown below. This section of the spectrum does not include the hydroxyl protons, –OH.

(i) What can be concluded about the two peaks?

(ii) Show the structure of the diol **F**. [5]

[Total: 14]

140

Chapter 6
Synoptic assessment

What is synoptic assessment?

Synoptic assessment emphasises your understanding and your application of the principles included in your chemistry course.

Part of your chemistry course is assessed using **synoptic questions**. These are written so that you can **draw together** knowledge, understanding and skills learned in **different parts** of AS and A2 Chemistry.

Synoptic assessment takes place mainly in **module exams** but part may be assessed in coursework.

What type of questions will be asked?

You will need to answer **two** main types of synoptic question.

You make the links between different areas of chemistry yourself. You choose and use the context for your answer.

1 **You make links** and **use connections** between different areas of chemistry. Some examples are given below.

Using examples drawn from different parts of your chemistry course:
- discuss the role of a lone pair in chemistry...
- discuss the chemistry of water...
- compare the common types of chemical bonding...

2 **You use ideas and skills** which **permeate chemistry**.

The context has been made for you. Often this will be a situation or will involve data that you will not have seen before.

This type of question will be more structured and you are unlikely to have as much choice in how you construct your answer.

Your task is to interpret any information using the 'big ideas' of chemistry.

Examples of themes that permeate chemistry are shown below:
- formulae, moles, equations and oxidation states
- chemical bonding and structure
- periodicity
- reaction rates, chemical equilibrium and enthalpy changes.

How will you gain synoptic skills?

A synoptic question may present you with a new piece of information.

You may be expected to calculate formulae from data, write correct formulae, balance equations and perform quantitative calculations using the mole concept, use knowledge of the Periodic Table to predict reactions of unfamiliar elements or compounds, etc.

Luckily chemistry is very much a synoptic subject. Throughout your study of chemistry, you apply many of the ideas and skills learnt during AS Chemistry or your GCSE course.

In studying A2 Chemistry, you will have been using synoptic skills naturally, probably without realising it. You certainly can make little real progress in chemistry without a sound grasp of concepts such as equations, the mole, structure and bonding!

Synoptic assessment

Worked synoptic sample question

Here, some of the big ideas of chemistry are being tested: formulae, bonding, redox, moles, equations. These will always appear on an exam paper with synoptic questions. You have nothing to fear from this type of question provided that your chemistry is sound.

Nice clues in the question. You are given the structure of sulfuric acid for a reason. You are being tested on whether you can apply hydrogen bonding to a new situation. This is not much different from water's H-bonding!

Just keep your nerve when writing equations. You can choose any reasonable examples.
Do think about ionic charges. Many students blindly assume that a sulfate ion has a 1– charge but how can this be so if we have H_2SO_4 with $2H^+$ ions?

In (i), you are just using oxidation number rules.
– this should be 2 easy marks.
(ii) is far harder.
You have to make sure that there is the same oxidation number change up and down. Here, each S goes down by 8 from +6 to – 2.
Each I goes up by 1 from –1 to 0 ... so we need 8 × I
Finally balance the H_2O and balance the H^+.

Structured type question

In its reactions, sulfuric acid, H_2SO_4, can behave as an acid, an oxidising agent and as a dehydrating agent.

The displayed formula of sulfuric acid is shown below.

$$\begin{array}{c} H-O \\ \end{array} \!\!\! \diagdown \!\!\! \begin{array}{c} S \end{array} \!\!\! \diagup\!\!\!\diagup \begin{array}{c} O \\ \end{array}$$
$$\begin{array}{c} H-O \end{array} \!\!\! \diagup \diagdown\!\!\!\diagdown \begin{array}{c} O \end{array}$$

(a) The boiling point of sulfuric acid is higher than expected.

   Suggest and explain why the boiling point of sulfuric acid is higher than expected.

   Sulfuric acid molecules form hydrogen bonds ✓

   hydrogen bonds break (on boiling) ✓ [2]

(b) Dilute sulfuric acid takes part in the typical acid reactions, reacting with carbonates, metals and alkalis.

   Write balanced equations for reactions of sulfuric acid with

   (i) an alkali $\quad 2NaOH + H_2SO_4 \longrightarrow Na_2SO_4 + 2H_2O$

   (ii) a metal $\quad Mg + H_2SO_4 \longrightarrow MgSO_4 + H_2$

   (iii) a carbonate $\quad MgCO_3 + H_2SO_4 \longrightarrow MgSO_4 + CO_2 + H_2O$ [3]

(c) Concentrated sulfuric acid oxidises some halide ions to form the halogen.

   The unbalanced equation below represents the oxidation of iodide ions by sulfuric acid.

   $H^+ + SO_4^{2-} + I^- \longrightarrow I_2 + H_2S + H_2O$

   (i) Determine the oxidation numbers of sulfur and iodine on either side of the equation.

   $SO_4^{2-} : +6 \longrightarrow H_2S: -2$ ✓

   $I^- : -1 \longrightarrow I_2 : 0$ ✓ [2]

   (ii) Balance the equation.

   $10H^+ + SO_4^{2-} + 8I^- \longrightarrow 4I_2 + H_2S + 4H_2O$ ✓ [1]

Synoptic assessment

Worked synoptic sample question (continued)

Here you are provided with information and you must use your chemical knowledge and understanding to solve the problem.

*Use the information – the whole question is about **dehydration** (loss of water) caused by sulfuric acid.*

Good advice is to look for the obvious. If your answer looks like some chemistry that you have never seen before, then you have probably made a mistake!

Key points here:

X is black. What do you know that is black?

Y must be made out of H, C, or O.

Z is hard but is still based on dehydration.

(d) Concentrated sulfuric acid dehydrates many organic compounds, forming water as one of the products.

For example, sulfuric acid dehydrates propan–1–ol by eliminating water to form propene.

$CH_3CH_2CH_2OH \longrightarrow CH_3CH=CH_2$

Three other examples are shown below.
- Sulfuric acid dehydrates sucrose, $C_{12}H_{22}O_{11}$, to form a black solid, **X**.
- Sulfuric acid dehydrates methanoic acid to form a gas, **Y**, with the same relative molecular mass as ethene.
- Sulfuric acid dehydrates ethane-1,2-diol to form a compound **Z** with a relative molecular mass of 88.0.

Suggest the identity of **X**, **Y** and **Z**. Write equations for each reaction and deduce the structural formula of compound **Z**.

X: C ✓ $C_{12}H_{22}O_{11} \longrightarrow 12C + 11H_2O$ ✓

Y: CO ✓ $HCOOH \longrightarrow CO + H_2O$ ✓

Z: $C_4H_8O_2$ ✓ $2C_2H_6O_2 \longrightarrow C_4H_8O_2 + 2H_2O$ ✓

Structure:

 ✓

[7]

[Total: 15]

Practice examination questions

1 Organic acids occur widely in nature.

(a) Butanoic acid, $CH_3(CH_2)_2COOH$, and compound **W** are straight-chain organic acids present in sweat.

(i) Compound **W** was analysed and was found to have the percentage composition by mass:

C, 66.7%; H, 11.1%; O, 22.2%. $M_r = 144.0$

Determine the molecular formula of compound **W** and suggest its structural formula.

(ii) Dogs can track humans from the odours in their sweat.

Sweat containing equal amounts of butanoic acid and compound **W** produces more butanoic acid vapour than vapour from compound **W**.

Suggest and explain a reason for this. [6]

Synoptic assessment

Practice examination questions (continued)

(b) Compound **X** is a straight chain organic acid. A chemist analysed a sample of acid **X** by the procedure below.

The chemist first prepared a 100 cm³ solution of **X** by dissolving 4.35 g of **X** in water.

In a titration, 10.00 cm³ 0.500 mol dm⁻³ NaOH were neutralised by exactly 8.50 cm³ of solution **X**.

(i) Calculate the pH of the NaOH(aq) used in the titration.
$K_w = 1.00 \times 10^{-14}$ mol² dm⁻⁶.

(ii) Use the results to calculate the molar mass of acid **X** and suggest its identity. [8]

[Total: 14]

2 Refer to data on Page 133 for this question.

Compounds **A** to **G** are isomers of $C_6H_{12}O_2$.

(a) Isomer **A**, $C_6H_{12}O_2$, is a neutral compound. Acid hydrolysis of **A** forms compounds **X** and **Y**.

X and **Y** can also both be formed from propanal by different redox reactions.

X has an absorption in its IR spectrum at 1750 cm⁻¹.

Deduce the structural formulae of **A**, **X** and **Y**. Give suitable reagents, in each case, for the formation of **X** and **Y** from propanal and state the role of the acid in the hydrolysis of **A**. [8]

(b) Isomers **B**, **C**, **D** and **E** are acidic.

B, **C** and **D** are structural isomers that also have optical isomers.

In its proton NMR spectrum, **E** has three singlets. Deduce the structural formulae **B**, **C**, **D** and **E**. [4]

(c) Isomer **F**, $C_6H_{12}O_2$, has the structural formula shown below, on which some of the protons have been labelled.

$$CH_3-\overset{\overset{\displaystyle O}{\|}}{C}-\overset{a}{CH_2}-CH_2-O-\overset{b}{CH_2}-CH_3$$

A proton NMR spectrum is obtained for **F**. Predict the chemical shift for the protons labelled *a* and *b*. Explain the splitting patterns arising from protons *a* and *b*. [6]

(d) Isomer **G**, $C_6H_{12}O_2$, contains six carbon atoms in a ring. It has an IR absorption at 3270 cm⁻¹ and shows only three peaks in its proton NMR spectrum. Deduce a structural formula for **G**. [2]

[Total: 20]

Practice examination answers

Chapter 1 Rates and equilibria

1 (a) (i) 1st order with respect to H_2O_2 ✓
Double concentration of H_2O_2, rate doubles ✓
1st order with respect to I^- ✓
7 times concentration of I^-, rate × 7 ✓
Zero order with respect to H^+ ✓
Double concentration of H^+, rate stays constant ✓

(ii) Rate = $k[H_2O_2][I^-]$ ✓
2.8×10^{-2} dm^3 mol^{-1} s^{-1} ✓✓ [9]

(b) (i) The slowest step of a multi-step process ✓

(ii) $H_2O_2 + I^- \longrightarrow H_2O + IO^-$ ✓ [2]

[Total: 11]

2 (a) (i) Rate = $k[A]^2[B]$ ✓

(ii) 3 ✓

(iii) 27 ✓

(iv) $k = 3.14 \times 10^{-3}$ dm^6 mol^{-2} s^{-1} ✓✓ [5]

(b) (i) H^+ is a catalyst ✓
H^+ appears in the rate equation but not the overall equation ✓

(ii) $CH_3COCH_3(aq) + H^+(aq) \longrightarrow [CH_3COHCH_3(aq)]^+$ ✓ [3]

[Total: 8]

3 (a) $K_c = \dfrac{[CH_3COOH][CH_3CH_2OH]}{[CH_3COOCH_2CH_3][H_2O]}$ ✓

$K_c = 0.26$ ✓; No units ✓ [3]

(b) (i) Equilibrium would move to the left ✓ to counteract the added CH_3CH_2OH ✓

(ii) No effect ✓ K_c only changes with temperature ✓ [4]

[Total: 7]

4 (a) (i) $[H_2(g)]$, 0.44 mol dm^{-3}; ✓ $[I_2(g)]$, 0.02 mol dm^{-3}; ✓
$[HI(g)]$, 0.32 mol dm^{-3} ✓

(ii) $K_c = \dfrac{[HI(g)]^2}{[H_2(g)][I_2(g)]}$ ✓ $= \dfrac{0.32^2}{0.44 \times 0.02} = 11.6$ ✓ no units ✓ [6]

(b) Equilibrium moves to left ✓; Forward reaction is exothermic (or the reverse reaction is endothermic) ✓ [2]

[Total: 8]

5 (a) Proton donor ✓ [1]

(b) $HCOOH + HNO_3 \rightleftharpoons HCOOH_2^+ + NO_3^-(aq)$ ✓
HCOOH is a base because it accepts a proton ✓ [2]

(c) (i) pH = 0.75 ✓

145

Practice examination answers

(ii) $K_w = [H^+(aq)][OH^-(aq)]$ ✓
$[H^+(aq)] = 1.0 \times 10^{-14}/0.372 = 2.69 \times 10^{-14}$ mol dm^{-3} ✓ pH = 13.57 ✓

(iii) $K_a = \dfrac{[H^+(aq)] [HCOO^-(aq)]}{[HCOOH(aq)]} = \dfrac{[H^+(aq)]^2}{[HCOOH(aq)]}$ ✓
$[H^+(aq)] = \sqrt{(1.6 \times 10^{-4} \times 0.263)} = 6.49 \times 10^{-3}$ mol dm^{-3} ✓ pH = 2.19 ✓

(iv) $[H^+(aq)] = K_a \times [HCOOH(aq)]/[HCOO^-(aq)]$ ✓
$[H^+(aq)] = 1.6 \times 10^{-4} \times 0.255/0.325 = 1.26 \times 10^{-4}$ mol dm^{-3} ✓
pH = 3.90 ✓ [10]

[Total: 13]

6 (a) (i) A strong acid completely dissociates to donate protons.
A weak acid partially dissociates to donate protons. ✓
(ii) Proton acceptor ✓ [2]

(b) (i) $[OH^-(aq)] = 1.50 \times 10^{-3} \times 1000/25 = 0.060$ mol dm^{-3} ✓
$K_w = [H^+(aq)][OH^-(aq)]$ ✓
$[H^+(aq)] = 1.0 \times 10^{-14}/0.060 = 1.67 \times 10^{-13}$ mol dm^{-3} ✓ pH = 12.78 ✓

(ii) Amount of HCl added = 0.0125 mol ✓
Amount of HCl remaining = $0.0125 - 1.50 \times 10^{-3} = 0.011$ mol ✓
[HCl] = $0.011 \times 1000/75 = 0.147$ mol dm^{-3} ✓ pH = 0.83 ✓ [8]

(c) $[H^+(aq)] = 10^{-pH} = 0.0302$ mol dm^{-3} ✓
amount of HCl = $0.0302 \times 25/1000 = 7.55 \times 10^{-4}$ mol ✓
$[H^+(aq)]$ in diluted solution = $7.55 \times 10^{-4} \times 1000/40 = 0.0189$ mol dm^{-3} ✓
pH = 1.72 ✓ [4]

[Total: 14]

Chapter 2 Energy changes in chemistry

1 (a) (i)

definition	letter
1st ionisation energy of potassium	B
1st electron affinity of bromine	F ✓
the enthalpy of atomisation of potassium	C
the enthalpy of atomisation of bromine	A ✓
the lattice enthalpy of potassium bromide	E
the enthalpy of formation of potassium bromide	D ✓

(ii) $-394 - (89 + 419 + 112 - 325)$ ✓ $= -689$ kJ mol^{-1} ✓ [5]

(b) (i) KI is less exothermic because I$^-$ has a smaller charge density ✓ and attraction between K$^+$ and I$^-$ ions is weaker ✓

(ii) CaBr$_2$ is more exothermic because Ca^{2+} has a greater charge density ✓ and attraction between Ca^{2+} and Br$^-$ ions is stronger ✓ [4]

[Total: 9]

2 (a) (i) Mg^{2+}(g) + 2Cl$^-$(g) ⟶ MgCl$_2$(s) ✓
(ii) Mg^{2+}(g) + aq ⟶ Mg^{2+}(aq) ✓
(iii) MgCl$_2$(s) + aq ⟶ Mg^{2+}(aq) + 2Cl$^-$(aq) ✓ [3]

(b) ΔH(solution) = $-$(lattice enthalpy) + Σ(enthalpy changes of hydration) ✓
= $-(-2526) + (-1891 + 2 \times -384)$ ✓ = -133 kJ mol^{-1} ✓ [3]

[Total: 6]

Practice examination answers

3 $\Delta H_r = \Sigma \Delta H_f(\text{products}) - \Sigma \Delta H_f(\text{reactants})$
   $= (-1676) - (-602)$ ✓
   $= -1074$ kJ mol^{-1} ✓
   $\Delta S_r = \Sigma S (\text{products}) - \Sigma S (\text{reactants})$
   $= [(2 \times 81) + 51] - [27 + (2 \times 28)]$ ✓
   $= 130$ J K mol^{-1} ✓ $= 0.13$ kJ K^{-1} mol^{-1} ✓
   $\Delta G = \Delta H - T\Delta S$ ✓
   $\Delta G = -1074 - 298 \times 0.13 = -1113$ kJ mol^{-1} ✓
   Reaction is feasible at 298 K because $\Delta G < 0$ ✓

   [Total: 8]

4 (a) 298 K; [Ni^{2+}(aq)] 1 mol dm^{-3} ✓; [Fe^{2+}(aq)] and [Fe^{3+}(aq)] are both 1 mol dm^{-3} ✓ [2]
   (b) It allows ions to flow between half-cells ✓ [1]
   (c) (i) 1.02 V ✓
       (ii) Electrons flow along wire from nickel half-cell ✓
       (iii) Ni is – electrode; Pt is + electrode ✓ [3]
   (d) (i) Ni ⟶ Ni^{2+} + 2e$^-$ ✓
           Fe^{3+} + e$^-$ ⟶ Fe^{2+} ✓
       (ii) 2Fe^{3+} + Ni ⟶ 2Fe^{2+} + Ni^{2+} ✓ [3]
   (e) In Ni redox equilibrium, increase in [Ni^{2+}(aq)] shifts equilibrium to the right. Electrons are less available and electrode potential becomes less negative. ✓
       Difference in cell potentials becomes less and cell potential will decrease. ✓ [2]

   [Total: 11]

5 (a) (i) Mn^{3+}(aq) ✓
       (ii) Mn^{3+}(aq) ✓ I$^-$ must provide electrons and have the more negative $E^\ominus$ (+0.54). ✓ The more positive $E^\ominus$ system (+1.49 V) will gain electrons. ✓ [4]
   (b) 0.66 V ✓
       4V^{2+}(aq) + O$_2$(g) + 2H$_2$O(l) ⟶ 4OH$^-$(aq) + 4V^{3+}(aq) ✓✓ [3]

   [Total: 7]

Chapter 3 The Periodic Table

1 (a) MgO, Al$_2$O$_3$, SiO$_2$; ✓ P$_4$O$_{10}$, SO$_3$ ✓ [2]
   (b) The oxidation number matches the number of outer shell electrons involved in bonding to oxygen. ✓ [1]
   (c) MgO(s) + H$_2$O(l) ⟶ Mg(OH)$_2$(aq) ✓
       Al$_2$O$_3$ and SiO$_2$ are insoluble ✓
       P$_4$O$_{10}$(s) + 6H$_2$O(l) ⟶ 4H$_3$PO$_4$(aq) ✓
       SO$_3$(l) + H$_2$O(l) ⟶ H$_2$SO$_4$(aq) ✓ [4]
   (d) Giant ionic oxides form alkaline solutions ✓
       Giant covalent oxides are insoluble ✓
       Simple molecular oxides form acidic solutions ✓ [3]

   [Total: 10]

2 (a) Ni: 1s^{2}2s^{2}2s^{6}3s^{2}3p^{6}3d^{8}4s^2 ✓; Ni^{2+}: 1s^{2}2s^{2}2s^{6}3s^{2}3p^{6}3d^8 ✓ [2]
   (b) A d-block element has its highest energy electron in a d sub-shell. ✓
       A transition element has at least one ion with a partially filled d sub-shell. ✓ [2]

Practice examination answers

    (c) (i) Lone pair of electrons on oxygen atom ✓
       (ii) Ion $[Ni(H_2O)_6]^{2+}$ ✓
       (iii) Covalent bonding in H_2O ✓ and coordinate bonding from water ligands to the Ni^{2+} ion. ✓
       (iv) Octahedral ✓ ; 90° ✓   [6]

    (d) (i) $Ni(OH)_2$ or $Ni(OH)_2(H_2O)_4$ ✓
         $Ni^{2+}(aq) + 2OH^-(aq) \longrightarrow Ni(OH)_2(s)$ ✓
         or $[Ni(H_2O)_6]^{2+}(aq) + 2OH^-(aq) \longrightarrow Ni(OH)_2(H_2O)_4(s) + 2H_2O(l)$
      (ii) $[NiCl_4]^{2-}$ ✓
      (iii) $[Ni(H_2O)_6]^{2+}(aq) + 6NH_3(aq) \longrightarrow [Ni(NH_3)_6]^{2+}(aq) + 6H_2O(l)$ ✓
         ligand substitution ✓   [5]

                                                                      [Total: 15]

3 (a) (i) +6 ✓
      (ii) +1 ✓
      (iii) +4 ✓   [3]
  (b) $2Cr^{3+} + 10OH^- + 3H_2O_2 \longrightarrow 2CrO_4^{2-} + 8H_2O$ ✓   [1]
  (c) Vanadium as V_2O_5 in the contact process for the manufacture SO_3 for H_2SO_4 ✓
      Iron in the Haber process for the manufacture of NH_3 ✓   [2]

                                                                        [Total: 6]

4 (a) (i) +7 ✓
      (ii) $5H_2O_2 + 2MnO_4^- + 6H^+ \longrightarrow 5O_2 + 8H_2O + 2Mn^{2+}$ ✓   [2]
  (b) Reducing agent ✓   [1]
  (c) Amount of $KMnO_4 = 0.0150 \times \dfrac{23.80}{1000} = 3.57 \times 10^{-4}$ mol ✓
      Amount of H_2O_2 that reacted $= 2.5 \times 3.57 \times 10^{-4} = 8.925 \times 10^{-4}$ mol ✓
      Concentration of $H_2O_2 = \dfrac{1000}{25.0} \times 8.925 \times 10^{-4} = 0.0357$ mol ✓   [3]

                                                                        [Total: 6]

5 (a) (i) A coordinate (dative covalent bond) forms ✓ between a lone pair on a ligand and the central metal ion. ✓
      (ii) Bidentate ligand ✓   [3]
  (b) (i) +3 ✓
      (ii) $1s^2 2s^2 2s^6 3s^2 3p^6 3d^6$ ✓   [2]
  (c) (i) 6 ✓
      (ii) Octahedral ✓
      (iii) Add an excess of 1,2-diaminoethane or $NH_2CH_2CH_2NH_2$ ✓
         $[Ni(H_2O)_6]^{2+} + 3NH_2CH_2CH_2NH_2 \rightarrow [Ni(NH_2CH_2CH_2NH_2)_3]^{2+} + 6H_2O$ ✓   [4]

                                                                        [Total: 9]

Practice examination answers

Chapter 4 Chemistry of organic functional groups

1. (a) C : H : O = 66.7/12 : 11.1/1 : 22.2/16 ✓
   Molecular formula = C_4H_8O ✓ [2]
   (b) $CH_3CH_2COCH_3$ ✓; $CH_3CH_2CH_2CHO$ ✓; $(CH_3)_2CHCHO$ ✓ [3]
   (c) 2,4-dinitrophenylhydrazine ✓ orange precipitate ✓ [2]
   (d) Tollens' reagent ✓. Reacts with butanal and methylpropanal ✓
   Silver mirror forms ✓ (or $H_2SO_4/K_2Cr_2O_7$; from orange to green) [3]
   (e) Product: $CH_3CH_2CH(OH)CH_3$ ✓ formed from $CH_3CH_2COCH_3$ ✓ [2]
   [Total: 12]

2. (a) (i) Ethyl propanoate ✓
   (ii) E: CH_3CH_2OH ✓
   F: $CH_3CH_2COO^-Na^+$ ✓✓ (1 mark if CH_3CH_2COOH) [4]
   (b) (i)
   HCOO—CH_2
            |
   HCOO—CH_2 ✓

   (ii) Concentrated H_2SO_4 ✓ [2]

   (c)
   $3CH_3COOH$ + HO—CH_2 / HO—CH / HO—CH_2 ✓✓ → CH_3COO—CH_2 / CH_3COO—CH / CH_3COO—CH_2 + $3H_2O$ ✓ [3]
   [Total: 9]

3. (a) (i) Molar ratio C:H = 90.56/12 : 9.44/1 ✓
   Empirical formula = C_4H_5 ✓
   Molecular formula = C_8H_{10} ✓
   (ii) [four isomers of dimethylbenzene / ethylbenzene structures shown] ✓✓✓✓ [7]
   (b) (i) 1,4-dimethylbenzene ✓
   (ii) [structure with CH_3, NO_2, CH_3 on ring] ✓ [2]
   [Total: 9]

4. (a) (i) Electrophilic substitution ✓
   (ii) Concentrated HNO_3 and concentrated H_2SO_4 ✓ [2]
   (b) (i) Tin and concentrated HCl ✓
   (ii)
   ✓✓ [3]
   [Total: 5]

Practice examination answers

5 (a) Compound A: $CH_3CH_2CH_2COOH$ ✓
    Compound B: $(CH_3)_2CHOH$ ✓ [2]
  (b) Step 1: $H_2SO_4/K_2Cr_2O_7$ ✓ reflux ✓ oxidation ✓
    Step 2: $NaBH_4$ ✓ with H_2O or ethanol ✓ reduction ✓
    Step 3: Conc. H_2SO_4 catalyst ✓ reflux ✓ esterification ✓ [9]
                                                            [Total: 11]

Chapter 5 Polymers, analysis and synthesis

1 (a) $(CH_3)_2CH$ ✓ [1]
  (b) (i) [structure of dipeptide with peptide bond circled] ✓
      (ii) [structure of alanine: $H_2N-CH(CH_3)-COOH$] ✓
      (iii) [structure of dipeptide of valine and alanine] ✓ [3]
  (c) (i) A chiral carbon atom has **four** different groups attached to it. ✓ It provides two possible forms that are non-superimposable mirror images of one another. ✓
      (ii) [two mirror-image structures of valine with COOH, H, $(CH_3)_2CH$, NH_2] ✓✓
          Bond angle = 109.5° ✓ [5]
  (d) [three structures of valine at pH = 2, pH = 7, pH = 12] ✓✓✓ [3]
  (e) [structure of serine: $H_2N-CH(CH_2OH)-COOH$] ✓ [1]
                                                            [Total: 13]

Practice examination answers

2 (a) 2-hydroxypropanoic acid ✓ [1]

(b) (i) A chiral carbon with four different groups attached ✓

(ii) Optical isomers are non-superimposable mirror images ✓

(iii)

```
      COOH              COOH
       |                 |
       C⋯⋯H         H⋯⋯C
      / \              / \
   H₃C   OH          HO   CH₃     ✓✓
```
[4]

[Total: 5]

3 (a) (i)

```
    H         H
     \       /
      C = C
     /       \
   H₃C        H    ✓
```

(ii) Addition ✓ [2]

(b) (i)

```
          H   H                    O               O
          |   |                    ‖               ‖
   H₂N — C — C — NH₂        HO — C —⟨ ⟩— C — OH
          |   |
         CH₃ CH₃                                   ✓✓
```

(ii) Condensation ✓ [3]

(c) Polymer **D** has a polar CONH group. Polymer **C** is non-polar. ✓
Polar group in polymer **D** will attract water or hydroxide ions. ✓ [2]

[Total: 7]

4 (a) (i) 4 ✓

(ii) C_2H_5 ✓
The CH_2 group at δ = 1.63 ppm is split into a quartet ✓ by 3 protons on an adjacent carbon. ✓
The CH_3 group at δ = 0.90 ppm is split into a triplet ✓ by 2 protons on an adjacent carbon. ✓

(iii) A CH_3 group with no protons on an adjacent carbon. ✓

(iv) Two equivalent OH protons as the absorption disappears with D_2O. ✓

(v) $CH_3CH_2C(OH)_2CH_3$ ✓ [9]

(b) (i) There are 8H atoms with a peak ratio of 1:3. The peak at δ = 3.64 ppm is from a CH_2–O ✓ with no protons on an adjacent carbon atom ✓
The peak at δ = 1.24 ppm is from two equivalent CH_3 groups ✓ with no protons on an adjacent carbon atom. ✓

(ii) $(CH_3)_2C(OH)CH_2OH$ ✓ [5]

[Total: 14]

151

Practice examination answers

Chapter 6 Synoptic assessment

1 (a) (i) $C : H : O = \dfrac{66.7}{12} : \dfrac{11.1}{1} : \dfrac{22.2}{16}$ ✓

$= 5.56 : 11.1 : 1.39 = 4 : 8 : 1$
empirical formula = C_4H_8O ✓
$48 + 8 + 16 = 72$ which is half of M_r
Therefore molecular formula = $C_8H_{16}O_2$ ✓
Structural formula = $CH_3(CH_2)_6COOH$ ✓

(ii) compound **W** has a longer carbon chain and contains more electrons ✓
compound **W** has more van der Waals' forces ✓ [6]

(b) (i) $[H^+(aq)] = \dfrac{K_w}{[OH^-(aq)]}$ ✓ $= \dfrac{1.00 \times 10^{-14}}{0.500}$

$= 2.00 \times 10^{-14}$ mol dm^{-3} ✓
$pH = -\log[H^+(aq)] = -\log(2.00 \times 10^{-14}) = 13.70$ ✓

(ii) moles NaOH in 10.00 cm^3 = moles NaOH = 0.005 00 mol ✓
moles **A** in 8.50 cm^3 = moles NaOH = 0.005 00 mol ✓
moles **A** in 100 cm^3 = $\dfrac{0.005\,00 \times 100}{8.50}$ = 0.0588 mol ✓

molar mass of **A** = $\dfrac{4.35}{0.0588}$ = 74.0 g mol^{-1} ✓

Therefore **A** is propanoic acid / CH_3CH_2COOH ✓ [8]

[Total: 14]

2 (a) **A** = $CH_3CH_2COOCH_2CH_2CH_3$ ✓
X = CH_3CH_2COOH ✓
Y = $CH_3CH_2CH_2OH$ ✓
For propanal ⟶ **X**, H_2SO_4 ✓ and $K_2Cr_2O_7$ ✓
For propanal ⟶ **Y**, $NaBH_4$ ✓ in water ✓
Acid is a catalyst ✓ [8]

(b) **B**, **C** and **D**
$CH_3CH_2CH_2CH(CH_3)COOH$ ✓
$CH_3CH_2CH(CH_3)CH_2COOH$ ✓
$(CH_3)_2CHCH(CH_3)COOH$ ✓
E: $(CH_3)_3CCH_2COOH$ ✓ [4]

(c) a: δ = 3.3–4.3 ppm ✓; y: δ = 2.0–2.9 ppm ✓
a: splitting will be a quartet ✓ as there is an adjacent CH_3 ✓
b: splitting will be a triplet ✓ as there is an adjacent CH_2 ✓ [6]

(d) [cyclohexane ring structure with OH groups at 1 and 4 positions]

✓ (structure) ✓ (2 × OH groups at correct positions) [2]

[Total: 20]

Notes

Notes

Notes

Notes

Notes

Periodic Table

The Periodic Table

Key:
- relative atomic mass (top)
- atomic symbol
- atomic number (bottom)
- name

Example: 1.0 / H / 1 / Hydrogen

1	2											3	4	5	6	7	0
																	4.0 He 2 Helium
6.9 Li 3 Lithium	9.0 Be 4 Beryllium											10.8 B 5 Boron	12.0 C 6 Carbon	14.0 N 7 Nitrogen	16.0 O 8 Oxygen	19.0 F 9 Fluorine	20.2 Ne 10 Neon
23.0 Na 11 Sodium	24.3 Mg 12 Magnesium											27.0 Al 13 Aluminium	28.1 Si 14 Silicon	31.0 P 15 Phosphorus	32.1 S 16 Sulfur	35.5 Cl 17 Chlorine	39.9 Ar 18 Argon
39.1 K 19 Potassium	40.1 Ca 20 Calcium	45.0 Sc 21 Scandium	47.9 Ti 22 Titanium	50.9 V 23 Vanadium	52.0 Cr 24 Chromium	54.9 Mn 25 Manganese	55.8 Fe 26 Iron	58.9 Co 27 Cobalt	58.7 Ni 28 Nickel	63.5 Cu 29 Copper	65.4 Zn 30 Zinc	69.7 Ga 31 Gallium	72.6 Ge 32 Germanium	74.9 As 33 Arsenic	79.0 Se 34 Selenium	79.9 Br 35 Bromine	83.8 Kr 36 Krypton
85.5 Rb 37 Rubidium	87.6 Sr 38 Strontium	88.9 Y 39 Yttrium	91.2 Zr 40 Zirconium	92.9 Nb 41 Niobium	95.9 Mo 42 Molybdenum	[98] Tc 43 Technetium	101.1 Ru 44 Ruthenium	102.9 Rh 45 Rhodium	106.4 Pd 46 Palladium	107.9 Ag 47 Silver	112.4 Cd 48 Cadmium	114.8 In 49 Indium	118.7 Sn 50 Tin	121.8 Sb 51 Antimony	127.6 Te 52 Tellurium	126.9 I 53 Iodine	131.3 Xe 54 Xenon
132.9 Cs 55 Caesium	137.3 Ba 56 Barium	138.9 La* 57 Lanthanum	178.5 Hf 72 Hafnium	180.9 Ta 73 Tantalum	183.8 W 74 Tungsten	186.2 Re 75 Rhenium	190.2 Os 76 Osmium	192.2 Ir 77 Iridium	195.1 Pt 78 Platinum	197.0 Au 79 Gold	200.6 Hg 80 Mercury	204.4 Tl 81 Thallium	207.2 Pb 82 Lead	209.0 Bi 83 Bismuth	[209] Po 84 Polonium	[210] At 85 Astatine	[222] Rn 86 Radon
[223] Fr 87 Francium	[226] Ra 88 Radium	[227] Ac* 89 Actinium	[261] Rf 104 Rutherfordium	[262] Db 105 Dubnium	[266] Sg 106 Seaborgium	[264] Bh 107 Bohrium	[277] Hs 108 Hassium	[268] Mt 109 Meitnerium	[271] Ds 110 Darmstadtium	[272] Rg 111 Roentgenium							

Elements with atomic numbers 112–116 have been reported but not fully authenticated

lanthanides

140.1 Ce 58 Cerium	140.9 Pr 59 Praseodymium	144.2 Nd 60 Neodymium	144.9 Pm 61 Promethium	150.4 Sm 62 Samarium	152.0 Eu 63 Europium	157.2 Gd 64 Gadolinium	158.9 Tb 65 Terbium	162.5 Dy 66 Dysprosium	164.9 Ho 67 Holmium	167.3 Er 68 Erbium	168.9 Tm 69 Thulium	173.0 Yb 70 Ytterbium	175.0 Lu 71 Lutetium

actinides

232.0 Th 90 Thorium	[231] Pa 91 Protactinium	238.1 U 92 Uranium	[237] Np 93 Neptunium	[242] Pu 94 Plutonium	[243] Am 95 Americium	[247] Cm 96 Curium	[245] Bk 97 Berkelium	[251] Cf 98 Californium	[254] Es 99 Einsteinium	[253] Fm 100 Fermium	[256] Md 101 Mendelevium	[254] No 102 Nobelium	[257] Lr 103 Lawrencium

Index

Absorption 69–70
Acid anhydrides 101–102
Acid dissociation constant, K_a 26
Acid–base pairs 24–25
Acid–base properties 116–117
Acids 117
Acids and bases 62
Acyl chlorides 100–101
Acylation 100–102, 108
Additional polymerisation 120–121
Addition–elimination reactions 101
Alcohols 127
Aldehydes 90–93
Aliphatic amines 109, 110
Aliphatic synthetic routes 136
Amides 112
Amines 109–112
Amino acids 116–118
Analysis 126–129
Aqueous bases 82–83
Arenas 103–108
Aromatic amines 109, 110
Aromatic organic compounds 103–104
Aromatic synthetic routes 137
Aspirin synthesis 102
Atom economy 23
Autocatalysis 79

Bases 110, 117
Benzene 104–105, 137
Biodiesel 99
Boiling points 60–61, 64
Bond enthalpy 43–44
Born–Haber cycle 39–41
Brønsted–Lowry acids and bases 24–25
Buffer solutions 33–35
Butane 128

Carbonyl compounds 90–92
Carboxyl group 94
Carboxylates 95
Carboxylic acids 94–96, 127
Catalysts 16, 64, 77–79
Cell diagrams 51
Cell reactions 50–51
Chirality 88
Chromatography 125
Chromium 76, 78
cis-trans isomerism 68

Cobalt 76
Colorimetry 70
Colour 68, 69
Column chromatography 125
Complex ions 64, 68, 71–73
Compounds 64
Condensation polymerisation 121–122

d-block elements 63
Delocalised structure 104–105
Density 64
Dissociation of acids 25–26
Drug synthesis 88

Electrochemical cells 48–53, 52
Electrochemical series 50
Electron configuration 62
Electrophilic substitution 107
Enthalpy changes 39–44
Entropy changes 45–46
Equilibrium constants, K_c 19–22
Equilibrium law 19–20
Equilibrium, rates and equilibria 13–35
Ester 127
Esterification 96
Esters 97–99

Fats 98–99
Fehling's solution 93
Formation of ions 64–65
Formulae 60
Fragment ions 128
Free energy 46
Fuel cells 52–53
Functional groups 87–89, 103

Gas–Liquid chromatography 125
Giant structures 61

Haemoglobin 72–73
Halogenoalkanes 111
Heterogeneous catalysis 77–78
Homogeneous catalysis 78–79
Hydrogenation 106
Hydrolysis 97–98, 99, 118

Infra-red absorptions 126
Infra-red spectra 126–127
Infra-red spectroscopy 126

Initial rates method 14–15
Ionisation of water 29–30
Isoelectric points 116–117
Isomerism 87–89

Kekulé structure 104
Ketones 90–93
Kevlar 122

Lattice enthalpy 39–41
Ligands 66–67, 71–73

Mass spectrometry 127–128
Mass spectrum 128–129
Melting points 64
Metal aqua-ions 80–83
Metal hydroxides 81–82
Metal oxides 59–60
Molecular ions 127
Monomers 120, 122, 123

Nitration 107
Nitriles 110
Nitroarenes 110
NMR carbon-13 131–132
NMR proton 133
Non-metal oxides 60, 61
Nuclear magnetic resonance (NMR) spectroscopy 130–135
Nucleic acids 120–124
Nucleophiles 100–101
Nucleophilic addition 91–92

Oils 98–99
Optical isomers 117
Orders of reaction 13–16
Organic synthesis 100
Organic synthetic routes 136–137
Oxidation states 65

Period 3 elements 59–62
Periodic Table 59–83
pH 27–30
Polarity 94, 109
Polyalkenes 123
Polyamides 121–122
Polyesters 122–123
Polymers 120–124
Polypeptides 118, 124
Precipitation reactions 80–81
Proteins 118, 120–124

159

Index

Rate constant, k 15–16
Rate equation 13–16
Rate-determining step 17–18
Reactions 107–108
Reactivity 64
Redox reactions 54–55, 74–76

Solvents 135
Spontaneous processes 46
Standard cell potentials 50–51, 54–55

Standard electrode potentials 49–50
Stereoisomerism 67–68, 87–88
Strong acids 26, 27–28, 31
Strong alkalis 30, 31
Structural isomerism 87–88
Synoptic assessment 141

Thermochemical evidence 105–106
Titration calculations 32–33
Titration curves 31–32

Transition elements 63–65, 66–70, 77
Transition metal ions 74–76, 79

Vanadium 75, 78

Weak acids 25, 28–29, 31, 33
Weak alkalis 31

Zwitterions 116–117